Long Problems

Long Problems

CLIMATE CHANGE AND THE CHALLENGE
OF GOVERNING ACROSS TIME

THOMAS HALE

PRINCETON UNIVERSITY PRESS
PRINCETON & OXFORD

Copyright © 2024 by Princeton University Press

Princeton University Press is committed to the protection of copyright and the intellectual property our authors entrust to us. Copyright promotes the progress and integrity of knowledge. Thank you for supporting free speech and the global exchange of ideas by purchasing an authorized edition of this book. If you wish to reproduce or distribute any part of it in any form, please obtain permission.

Requests for permission to reproduce material from this work should be sent to permissions@press.princeton.edu

Published by Princeton University Press
41 William Street, Princeton, New Jersey 08540
99 Banbury Road, Oxford OX2 6JX

press.princeton.edu

All Rights Reserved

Library of Congress Cataloging-in-Publication Data

Names: Hale, Thomas (Thomas Nathan), author.
Title: Long problems : climate change and the challenge of governing across time / Thomas Hale.
Description: Princeton : Princeton University Press, [2024] | Includes bibliographical references and index.
Identifiers: LCCN 2023030326 (print) | LCCN 2023030327 (ebook) | ISBN 9780691238128 (hardback) | ISBN 9780691238135 (ebook)
Subjects: LCSH: Climatic changes—Government policy. | Climatic changes—Economic aspects. | Climatic changes—International cooperation.
Classification: LCC QC903 .H215 2024 (print) | LCC QC903 (ebook) | DDC 363.7/06—dc23/eng/20231012
LC record available at https://lccn.loc.gov/2023030326
LC ebook record available at https://lccn.loc.gov/2023030327

British Library Cataloging-in-Publication Data is available

Editorial: Bridget Flannery-McCoy and Alena Chekanov
Production Editorial: Natalie Baan
Jacket Design: Chris Ferrante
Production: Erin Suydam
Publicity: James Schneider and Kathryn Stevens
Copyeditor: Joyce H.-S. Li

This book has been composed in Arno Pro

Printed on acid-free paper. ∞

Printed in the United States of America

10 9 8 7 6 5 4 3 2 1

For my students

CONTENTS

Acknowledgments		ix
1	Long Problems	1
	A Carbon View of Politics	1
	The Argument in Brief	2
	Defining Problems and Their Length	5
	Why We Need to Govern Long Problems	12
	About This Book	15
2	Why Long Problems Are Hard to Govern	21
	Short-Termism: The Tyranny of the Present	22
	Shadow Interests: Decision-Takers Have No Power over Decision-Makers	29
	Institutional Lag: Dynamic Problem Structures	32
	The Early Action Paradox: Uncertainty, Salience, Obstructionism, and Muddling Through	37
3	Forward Action: Addressing the Early Action Paradox	43
	Information and Foresight: Making the Future Known and Salient	44
	Experimentalism: Overcoming Uncertainty	54
	Catalysts: Eroding Obstructionism	60
4	The Long View: Addressing Shadow Interests	70
	Representation: Voices for Future Generations	71
	Trusteeship: Insulation and Tying Our Hands	78
	Horizon-Shifting: Changing Preferences	90

5	Endurance and Adaptability: Addressing Institutional Lag	102
	Endurance: Goal-Setting	105
	Reflexive Governance: Updating Institutions	115
	Triggers and Reserves: Combining Durability and Adaptability	122
6	Studying Long Problems	128
	Taking the Period of Analysis Seriously	129
	Existing Tools	137
	Rates of Change	143
	Dynamic Problem Structure Chains	149
	Empirical Analysis of the Future	157
7	Governing Time	166
	An Institutional Agenda on Climate Change	169
	How Do We Get There: Climate Change and Institutional Change	177
	Learning to Fly	180

Appendix 1: Why We Face More Long Problems Now		183
	The Collision of Human and Planetary Timescales	184
	Science and Technology Allow Us to See the Future More Clearly and to Affect It More Decisively	187
	A Changing Normative Frame? The Growing Moral Shadow of the Future	191
Appendix 2: The Role of Acceleration		195
Notes		199
References		213
Index		233

ACKNOWLEDGMENTS

The idea for this book grew in part from a series of workshops organized by Robert O. Keohane at the Center for Advanced Study in the Behavioral Sciences at Stanford University. In 2016, Bob received the Balzan Prize for International Relations, History and Theory. This prestigious award came with a twist: the prize money was to be spent on a project to nurture a new generation of scholars. This requirement made the award even more appropriate for Bob, who, alongside his field-defining scholarship, has provided extraordinary mentorship to at least two generations of political scientists.

Under Bob's direction, the Balzan workshops aimed to support a cohort of younger scholars working on the comparative and international study of climate politics. These meetings in the Palo Alto hills served to incubate my papers on catalytic cooperation and, with coauthors Jeff Colgan and Jessica Green, on asset revaluation and existential politics. That work brought the temporal dimensions of climate politics to the center of my thinking. Although most of my work has looked at governing across space, I thank this explicitly intergenerational prize and the fantastic scholars who attended the workshops it supported for leading me to think about governance across time.

No book on this theme can avoid an enormous intellectual debt to the work of Paul Pierson, Alan Jacobs, and Jonathan Boston, each of whom has made a critical contribution to the study of politics and governance across time. Without them, my thinking on this subject would have remained much more superficial and fragmented. Generous reading and critique from Bob Keohane, David Victor, Steven Bernstein, Jonathan Boston, and several reviewers helped prune, temper, and hone the ideas further.

Outside my field, I am also grateful to the normative scholars who, from a range of philosophical positions, have made such a clear case for why, morally, we should take long problems seriously. My hope with this book is to answer their call by investigating how, practically, we can do so.

In pursuing this goal, I am deeply thankful to accompany the growing community of thinkers and doers who are working to weave long-term thinking and governance across society. From the United Nations down to cities like Oxford, in all spheres of life, dedicated individuals are working to make our societies capable of tackling long problems like climate change. Their experience and wisdom have inspired this book. I hope it strengthens their arms.

Whereas the Stanford workshops helped seed the idea of the book, two other environments helped it grow. My institutional home, the Blavatnik School of Government, gave me the intellectual courage and support to focus on what really matters. It is a privilege and a never-ending source of stimulation and challenge to work as part of such a mission-driven academic institution.

However, due to the COVID-19 pandemic, much of the actual writing of the book took place not in the office but in my living room. On this front, I am deeply grateful to my husband, Dr. Sam Webster, and to all our friends and family near and far for providing the "whole-person" nourishment all authors need.

Another happy outcome of the Stanford workshops was meeting Bridget Flannery McCoy at Princeton University Press. Bridget and her colleague Alena Chekanov have been essential sources of encouragement, guidance, and patience over the "long problem" of writing a book during a global pandemic. I thank them and the whole team at the press for making these ideas appear before you now.

Long Problems

1

Long Problems

A Carbon View of Politics

I am carbon. I sink into a swamp roamed by dinosaurs. For eons, rock and silt settle over me. Their weight slowly presses me tighter and tighter, the passing millennia compacting me into peat, then lignite, then hard coal. Above me, the surface transforms from epoch to epoch. Continents drift, seas rise and fall, ice freezes and melts, but I remain unchanged. After an eternity, *Homo sapiens* are born.

Now things move fast. You dig me out of a pit. I blacken your lungs. You throw me in a furnace and the work of ages burns in a flash. Now I am flying, shooting out of a brick stack and high into the air. Trillions and trillions of molecules follow, pouring from ground to sky faster than ever before. In a geologic instant I blanket the earth.

But in what for me is a mere moment, your societies change beyond recognition. I bring industry, altering how you live and work and even what you believe. Those of you who exploit me most effectively become the most powerful empires ever seen, with armies, companies, and languages that cover the globe. Life speeds up. Instead of counting time from planting to harvest, you now orient your hours around train tables, factory shifts, and telegraphs. New ideas, inequities, and struggles lead to nationalism, democratization, labor movements, communism, fascism, and the most violence the world has ever seen. These conflagrations do not slow your transformation but speed it further. The social contract is torn up and rewritten through movements and revolutions. Your struggles for independence create scores of new nations, cooperating and competing through a growing array of international bodies. This interdependence combines with new technologies to create a hyperconnected form of capitalism that speeds life up again. Communication becomes instantaneous, but the surge of digital information enables new forms of control as well as transparency. Politics becomes a fight between those who win or lose from these shifts. Superpowers rise and fall. Your numbers grow eightfold and you live longer and better than your ancestors could have imagined. Only in the last fraction of this geologic second do you grasp that I, the key to your transformation, carry also the threat to upend it.

I float serenely above your frenzy. In the few hundred years of human upheaval that followed the Industrial Revolution, my numbers have nearly doubled. Unlike you, I persist. The part of me that floated up in the puff of smoke from the world's first steam engine is still mostly there. Much of that bit of me will remain for centuries to come, outliving more than one hundred generations of the descendants of those who dug it out originally. The vast bulk of me that came after, the billions of tons of me that you pour into the sky each year, will last longer still.

My accumulation in the sky is already trapping so much of the sun's heat that the average temperature of the planet is more than a full degree Celsius warmer than it was when you started to burn me. This warming now triggers changes that could last far beyond the thousand years or so over which I will degrade. Plants and animals that go extinct will not re-evolve in that time. Forests that burn or dry into savanna or desert will not regrow in that time. Coral reefs that bleach and dissolve will not re-form in that time. Ice caps and glaciers will not refreeze and seas will not recede in that time.

You must wonder what will become of you during this next millennium and beyond of change. The planetary stability you evolved in is no more. You can model what will happen to me, but you struggle to imagine your own future. No one can say exactly what your technologies will allow, what you will value, how you will organize yourselves, or which interests will win and which will lose. But you do know that your chances now depend heavily on how well you can manage me and the climate change I cause. This is of course a question of how you manage yourselves—a question of politics—not just today or tomorrow but for as long as I and my consequences last.

The Argument in Brief

Problems like climate change unfold over the course of multiple human life spans. But our policy processes, the politics around them, and even the social science that tries to understand them do not match this time frame.

This temporal disconnect parallels the expansion of political problems across space. Over the modern era, globalization increased flows of money, goods, people, and ideas across borders. Transboundary policy issues like trade, financial flows, migration, and cross-border pollution gained salience, enmeshing domestic and international politics. Faced with this new category of issues, society created a growing system of intergovernmental and transnational governance, forming the vast apparatus of global governance we have today. In parallel, social scientists developed new theoretical constructs like international regimes, cooperation theory, interdependence, and network governance to understand (and to seek to influence and improve) this system. As the "object" of politics and governance expanded across space, so too did political dynamics, institutions, and theories.

This book argues that the expansion of political problems across time requires a similar shift. Just as the "widening" of political problems across national boundaries has led to profound shifts in how we understand, study, and approach politics and governance, so too does their "lengthening" across time horizons.

Of course, political problems have always unfolded over time. But in our current epoch, changes in technology and ecology are putting time at the heart of politics in an unprecedented way. Climate change—the "long emergency"[1]— shows this clearly, but the dilemma of governance across time appears in myriad other challenges: managing new technologies like artificial intelligence (AI) and gene editing, demographic shifts toward an older population, infrastructure investment and urban planning, and many others. Although the book proposes a way of understanding and governing long problems in general, most of its examples focus on climate change.

I define problem length as the temporal gap between a problem's causes and effects, and long problems as those whose causes and effects span more than one human generation. However, the book focuses less on conceptualizing long problems and more on understanding their implications for politics and governance. It seeks to answer three questions.

Why are long problems hard to govern? Short-termism, the time inconsistency of preferences, and uncertainty about the future are widely acknowledged, among other temporal vexations, as barriers to effective policymaking. The book digs deeper into the mechanisms underlying these ideas to define precisely how and under what conditions they block solutions. It develops a political economy analysis of long-term governance, offering a new conceptualization of the political and governance challenges long problems pose, focusing on three:

- **The early action paradox**: Action that affects outcomes must occur well in advance of those outcomes, but such early action is stymied by uncertainty, low salience, and obstructionism.

- **Shadow interests**: People in the future have no agency or ability to shape politics in the present; their interests are mere shadows in politics.
- **Institutional lag**: Institutions created to address the early phase of a long problem struggle to remain useful as the problem's structure develops over time.

These three concepts provide an analytically useful way of studying the various political dynamics that often get lumped into general reflections on the problem of short-termism.

How can we govern long problems? There is a long history of political thought on how to address political dilemmas over time. Scholars, policymakers, and advocates have proposed dozens of mechanisms that aim to change decision-makers' motives, incentives, and capacities or that constrain them in different ways.[2] This book does not put forward a new silver bullet solution. Rather, I scrutinize the range of existing and proposed mechanisms with a social scientific test: How and under what conditions can we expect them to succeed? This analysis, which forms the bulk of the book, is organized around the three problems identified above: the early action paradox, shadow interests, and institutional lag. For each, I scan a range of existing and proposed solutions and evaluate what conditions—for example, what distributions of preferences and power, what institutional settings, what political strategies, and so forth—would actually be needed for them to succeed. Throughout, I argue that ultimately, effective governance of long problems requires political strategies that change incentives in the present. The result is a set of arguments on the most promising ways to move toward better governance of long problems, including a proposal in the conclusion for an agenda of institutional reforms to tackle climate change.

How can we study long problems? Finally, the book reviews how social science concepts and theories already help us understand long problems, notes their limits, and outlines a research agenda on the politics and governance of time. Taking problem length seriously changes how we interpret core policy challenges and the politics around them. For example, we can see climate change less as a free-riding problem or distributional problem and more as a transition problem. Inequality is less a matter of simple redistribution and more a matter of creating economic and social structures that create incentives for equality. In this way, problem length focuses researchers' attention on different dynamics and causal mechanisms than those commonly emphasized. Of course, social scientists already possess a formidable toolkit of theoretical approaches, concepts, and methods to tackle temporal issues. These include historical institutionalism, path dependence, discount rates, transition studies, agent-based modeling, behavioral experiments, and many others. I briefly review such tools, highlighting their strengths and limitations, before arguing that social science needs to go further. I lay out a research agenda in three parts: studying rates of change as opposed to final outcomes, theorizing "problem structure" dynamically, and embracing empirical techniques that allow us to develop probabilistic knowledge about the future. The last of these proposes a significant epistemological shift in contemporary social science, arguing that too narrow a focus on identifying causality limits theory production because it truncates the object of study to the past.

Overall, the book makes the case that long problems like climate change require a fundamental rethinking of our political and governance strategies. Just as the expansion of "communities of fate"[3] across national boundaries has radically shifted political behavior, institutions, and thought, the long timescale of the most critical problems confronting society stands to remake the theory and practice of politics and governance.

This introductory chapter continues by defining political problems and demonstrating how their length, the temporal distance between a problem's causes and effects, is a key characteristic of all political challenges. It then discusses why we may encounter more long problems today than in the past, even though politics and society seem to have in some sense sped up. Appendixes 1 and 2 dig into these points in more detail for the interested reader. The chapter ends by locating the argument in contemporary debates and summarizing the remainder of the book.

Defining Problems and Their Length

Problem length is the time period over which the primary causes and effects of a problem unfold. To clarify this definition, it is important first to explain what is meant by a "problem" and how to think about its causes and effects. Defining political problems (or issues, or challenges; I use the terms interchangeably) is more difficult than it may seem. Although anyone, if asked, could likely rattle off a list of current challenges the world faces, the process through which problems come to be seen as such is complex. Even though we commonly refer to problems in broad terms as if they were singular—for example, climate, inequality, war—these issues are of course more accurately seen as amalgamations of different problems. For example, "climate" is

Element	Climate change	Inequality	Pandemic disease
Material or social facts	Concentration of GHGs	Gini coefficient	Pathogen prevalence and characteristics
Technical and scientific processes	Spectrometers, theory of the greenhouse effect	Income surveys, theories of economic distribution	Infection tests, germ theory
Social understandings and preferences	Perceptions of climate risk	Norms around equality	Fear of disease
Political narratives, policies, and institutions	Demands of a climate protester, goals of the Paris Agreement	Definition of poverty in a welfare program	Allocation of public health budget

(← More objective / More constructed →)

FIGURE 1.1. Elements of a political problem

composed of major subproblems like mitigating greenhouse gases (GHGs) or adapting to climate impacts, each with its own countless subdivisions.[4] Moreover, political actors often do not share a common definition of a problem.

These complex "objects" of governance are partially given by social or material realities and partially defined by the processes of understanding and governing them.[5] A rich conceptualization of both objective and socially constructed elements is important for defining problem length.

At root, a political problem is a certain understanding of a collection of social and/or material facts that provides a frame for political behavior. Social and material facts like the distribution of wealth or the concentration of GHGs can exist independent of politics, but how they come to matter necessarily depends on the technical, social, and ultimately political processes through which they are understood, emphasized, institutionalized, and acted on. In this way, we can understand a political problem as consisting of at least four related elements (figure 1.1):

- A set of *material and/or social facts* (e.g., the concentration of GHGs in the atmosphere, the level of inequality in a society, the emergence of a deadly pathogen)
- The *technical and scientific processes* through which those facts are understood (e.g., spectrometers and an understanding of the greenhouse effect, surveys of income levels and economic theories of income distribution, microscopes and germ theory)

- *Social understandings* of these material and social facts, including how actors believe they will affect their interests (e.g., perceptions of climate risks, normative understandings of equality, fears of disease)
- *Political narratives, policies, and institutions* through which actors seek to shape outcomes toward their interests (e.g., the demands of a climate protester, the way a welfare program defines need, the choice of how to allocate public health investments)

Whereas the elements toward the upper end of this list are largely determined outside of social or political processes, those toward the bottom of the list are fundamentally social and political constructions. Indeed, these more social and political elements are so important that they can largely determine the political dynamics around a problem, even flying in the face of objective social and material realities. For example, infamously, scientists intent on justifying white supremacy devised theories and identified empirical collections of facts that aim to support that view.[6] Understanding political problems in this way means that the definition of a given political problem—climate change, inequality, a pandemic—is often contested, no matter what "the facts" are. Solutions and responses to problems are of course even more contested.

Constructivism has limits, however. Leaders may talk down the risk of a deadly pandemic in order to seek political advantage, but even the most Orwellian narrative cannot change epidemiology. Similarly, efforts to deny the reality of climate change have little hope of altering atmospheric physics. These "objective" elements are important because, as political scientist Alan Jacobs, who studies why governments invest over the long term or not, puts it, "the very slowness of many social, economic, and physical processes imposes a temporal structure on the logic of government action."[7]

Climate change superbly demonstrates the complexities of defining political problems. At present, the basic material facts and scientific theories of a changing climate are widely understood. Our emissions are changing the makeup of the atmosphere and therefore rapidly raising the earth's average temperature, affecting numerous planetary systems and the human systems that depend on them.[8] But this understanding has been fiercely contested and disputed over the past thirty years, as interest groups have sought to shape our collective understanding of the problem in a way that suits their goals. Just as with tobacco or acid rain, industry groups that feared regulation invested heavily to problematize, cast doubt on, and dispute the science of climate change.[9]

But even where there is consensus on the material facts, there may not be consensus on the definition of the problem. In the early years of global climate cooperation, the basic facts were understood to create, essentially, a prevention problem similar to other environmental concerns. If we see the emissions that cause warming as the problem, the solution is clearly to reduce them. For most of the last thirty years, and still today, the preponderance of both scholars and policymakers have seen the mitigation challenge as a collective action problem among states—how to get countries to act given the incentive to free ride.[10] But others have advocated seeing mitigation instead as a transition problem,[11] a distributional problem,[12] a lock-in problem,[13] a technological innovation problem,[14] an asset revaluation problem,[15] or through other lenses. I return to these alternative understandings in chapter 6. For now, the important point is that even where there is consensus on the material facts and broad objectives, there is not necessarily agreement on the nature of the problem overall. Most sophisticated observers would likely suggest that different aspects of the problem can be better or worse understood via some combination of these different lenses.

Moreover, reducing emissions is now understood to be only one aspect of the climate problem. This may seem obvious, but it was not always the case. As climate impacts became better understood and mitigation lagged, vulnerable countries pushed to expand the understanding of the climate problem to include efforts to adapt to climate change.[16] After all, even if prevention is perhaps preferable to treatment, a weak prevention strategy and no plans for treatment is cold comfort to those most at risk. Some mitigation advocates initially resisted this move as an admission of defeat or even a slippery slope toward giving up on prevention, while rich countries feared it might emphasize increasing financial support for the most vulnerable. But as the material facts of climate impacts have become impossible to ignore, adaptation has become a mainstream pillar of climate governance.

More recently, as climate impacts have continued to intensify, vulnerable interests are pushing to expand the climate problem further to include not just prevention and adaptation but also liability and compensation. After all, climate change is already creating "loss and damage," as the issue is termed in the United Nations (UN) process, which cannot be adapted to.

In the future, we can likely expect the definition of the problem to continue to evolve. For example, many argue we must understand the climate problem to include deployment of negative emissions technologies to suck carbon back out of the atmosphere (indeed, many scenarios for reaching global temperature goals assume them) or even solar radiation management, reflecting the sun's energy back into space through aerosol sprays or other means of geoengineering. In a different vein, some groups advocate treating the climate problem like crimes against humanity, making "ecocide" a grave criminal offense equivalent to genocide.

Understanding how problems are defined matters because different definitions lead to different political implications. Defining climate as a collective action problem suggests one set of solutions; seeing it as a compensation problem provides another. Political scientists often describe the characteristics of a problem as the "problem structure" of a given issue[17] or, in the language of game theory, what type of "game" is being played. For example, problems can have more or less uncertainty, involve a large number of actors or only a few, or be characterized by greater or lesser alignment of interests.[18]

I argue that problem length—*defined as the temporal distance between the primary causes and effects of an issue*—is another critical but underappreciated dimension across which problems vary.[19] This definition links three concepts. First, causes can be understood as any of the background factors or dynamics that create or contribute to the four elements of a problem described above (figure 1.1). Similarly, effects are the outcomes of those elements.[20]

Second, we can define primacy as how directly and how significantly a cause is linked to an effect in a chain of causal relationships. For example, fossil fuel emissions are the primary cause of climate change because they have a direct effect on global temperatures and account for the bulk of global warming. The spread of industrialization, which led to a large increase in emissions, is less proximate but still significant. The technological breakthroughs or economic systems that allowed for industrialization are more distant still. On the other end of the causal chain, the change in global temperatures is a proximate contributor to droughts in some regions of the world, such as the Middle East. Such droughts are one of many contributing factors to economic and social disruptions in countries like Syria, which are in turn one factor increasing the risk of political violence and, ultimately, the civil war that broke out there in 2011. We would certainly not say that climate change "caused" the civil war in Syria, as it was neither necessary nor sufficient for that tragic outcome. But it has been identified as a background factor.[21] A challenge with long problems is that chains of cause and effect may be quite extended, increasing the number of intervening factors and allowing a multiplicity of processes to shape outcomes.[22] Although every problem can ultimately be linked in various ways to a wide array of processes, from an analytic standpoint it often makes most sense to weight the relatively proximate and significant causes more heavily.

Finally, we can define problem length as the temporal distance—measured in seconds, years, centuries, or millennia—between a problem's primary causes and effects. Climate change is obviously a long problem. The material fact of global warming is caused by the centuries-long accumulation of GHGs, especially carbon, in the atmosphere, which will continue to warm the planet for hundreds if not thousands of years after the world achieves net zero emissions, should that happen. A forest fire, in contrast, flares up suddenly, and its effects may disappear within a generation. Climate change is a relatively long problem. A forest fire is a relatively short one. Of course, if one probes the underlying causes of the forest fire, reaching beyond the proximate, one may find a link to a long problem like climate change. This example demonstrates how the processes through which problem definitions are constructed affects how long or short we consider them to be. Seen as just a one-off event, a forest fire is a short problem. Seen as a climate impact, it becomes a very long one. Like other elements of problem structure, length is therefore partially given exogenously and partially constructed.

It is important to distinguish long problems, defined in this way, from ongoing ones.[23] Many problems persist over time, perhaps even indefinitely, but this does not make them long problems. For example, every government needs to focus constantly on issues like providing medical care or educating the young. These tasks will extend as far into the future as we can imagine because there will be new people to care for and educate, but their primary causes and effects fall within a single generation. In other words, these are not long problems but ongoing shorter problems. Even here, though, note that different understandings of the problem imply different problem lengths. Seeing health care as an issue of treating immediate needs makes it a very short problem. Focusing instead on prevention creates a much longer temporal frame that includes factors like maternal and neonatal health and childhood nutrition. Similarly, improving social welfare is primarily seen as a question of redistribution between present generations. But research has shown that intergenerational factors like parents' educational attainment and even their childhood nutrition can shape their offspring's well-being. Analytically, it is important to distinguish problems that recur over and over again from those whose causes and effects stretch across long periods.

The political problems we confront run the gamut from short to long. As the examples in table 1.1 demonstrate, problems that span decades, centuries, or millennia are heterogenous: problems are long in different ways. For example, political scientist Paul Pierson identifies different examples of

Table 1.1. Examples of problems with different lengths

Problem	Causes	Effects	Temporal gap
Emergency response to natural disasters	Hurricanes, floods, fires, etc.	Loss of human life and welfare, property	Minutes, hours, days
Pandemic diseases (e.g., flu, coronavirus)	New/mutated pathogens	Loss of human life and welfare, reduction in economic activity	Weeks, months
Armed conflict	Political disputes	Loss of human life and welfare, destruction of physical capital	Weeks, months, years
Chronic diseases (e.g., cancer, heart disease)	Lifestyle, environmental conditions	Loss of human life and welfare, reduction in economic activity	Years, decades
Antimicrobial resistance	Overuse of antibiotics	Decreased efficacy of basic medicines	Decades
Protecting renewable natural resources (e.g., forests, fisheries)	Overuse	Resource depletion	Decades
Technology development	Investment in research and development and other innovation support	Increased productivity and growth, positive spillovers	Decades
Public debt (e.g., bonds)	Current funding needs	Future tax burden	Decades
Increasing human capital	Education and training	Productivity	Decades
Repairing the ozone layer	Ozone-depleting substances	Increased radiation	Decades
Geopolitical power transition	Changing economic and military capacities	Interstate conflict	Decades
Infrastructure (e.g., roads, bridges, dams)	Depreciation through time and usage	Reduced usability, economic impacts	Decades
Social mobility/ marginalization	Access to education and social and economic opportunities	Greater equality	Decades, centuries
Urban planning	Built environment	Lifestyle and transportation behaviors	Decades, centuries
Accumulation of debris in earth orbit	Growing use of satellites without disposal plans	Risk to satellites	Decades, centuries
Accumulation of microplastics in the food chain	Plastic use, disposal	Biodiversity, food systems reduced	Decades, centuries
Climate adaptation	Climate impacts	Environmental, social, and economic disruptions	Decades, centuries, milennia
Climate mitigation	Greenhouse gases	Temperature change	Centuries, millennia
Storage of radioactive waste	Power production	Health and environmental risks	Millennia

Note: See also Boston 2016, 109.

slow-moving causal processes that can create long problems.[24] Cumulative processes like urbanization, migration, literacy, or the spread of national identities tend to accrete gradually over time. In contrast, threshold effects may have cumulative causes, but their outcomes manifest rapidly, like discontent that slowly builds up but then explodes in a revolution. Perhaps the most difficult to analyze are multistage causal chains in which X leads to Y, but via a series of intermediate steps that each have their own logics and lengths.

Within each of these patterns, long problems may allocate costs and benefits differently across their span. Problems like climate change have, on average, present costs and future benefits. Taking on public debt, in contrast, involves paying future costs for present benefits.[25] Similarly, other problem features like irreversibility can tend to be associated with long problems (because their effects often take a long time to play out and so cannot be reversed quickly), but there are also irreversible problems that are not long (such as a radical technological breakthrough).

Long problems are a diverse group because problem length is only one of many dimensions across which problems vary. I do not argue that length is the only meaningful way to understand climate change, which scholars have discussed as a "super wicked" or "creeping" problem, or other long problems.[26] Certainly, a full understanding of any problem requires attention to characteristics beyond length. However, my focus here is to show how attention to this one characteristic, which seems quite intuitive prima facie, can in fact fundamentally reshape our understanding of politics.

Why We Need to Govern Long Problems

Why should we seek to understand and govern long problems? Perhaps the best metaphor comes from Geoffrey Vickers, a British polymath who shaped, and was shaped by, the upheavals of most of the twentieth century. In his 1970 *Value Systems and Social Process*, in a chapter titled "The End of Freefall," Vickers compares modern society to a person jumping off a tall building and, on the way down, remarking, "Well, I am fine so far."[27]

Human development is, like freefall, an exhilarating rush but one that needs direction if it is to end well. Vickers argued that if human civilization was to survive, "it will have to be controlled—that is, governed—on a scale and to a depth which we have as yet neither the political institutions to achieve nor the cultural attitudes to accept."[28] That is the challenge long problems pose to a society beginning to glimpse the ground below coming into view.

To be sure, governing across time is not a new aspiration. European monarchs still hear their subjects shout "Long live the king/queen!" In dynastic China, officials proclaimed the ruler should endure ten thousand years—meaning essentially forever—a phrase the Chinese Community Party also applied to Chairman Mao Zedong. In Nazi Germany, Adolf Hitler envisioned a thousand-year Reich. The 1789 American Constitution, like many of the written constitutions that have followed it, promises "to secure the Blessings of Liberty to ourselves and our Posterity." And the Charter of the United Nations begins by pledging "to save succeeding generations from the scourge of war." Indeed, the very idea of governance seems to require the durability of political order.

The difference now is that the objects of governance are increasingly in the future.[29] There are at least three reasons why long problems are increasingly prevalent. I explore these drivers more fully in appendix 1.

First, the scale and speed of development increasingly brings human systems into contact with planetary systems—like the carbon cycle—that operate on radically different timescales. As the economy has expanded, humanity's footprint on the planet has begun to alter the earth's fundamental geophysical, chemical, and biological systems. Human societies have of course ravaged parts of the world before, denuding Easter Island, killing off the megafauna of Australia or Madagascar, or fencing the American Great Plains. But around the middle of the twentieth century, we began to pass a threshold between localized and system-wide impacts, a period termed the "Great Acceleration."[30] Many describe the present epoch as the Anthropocene because humans are now the primary driver of planetary outcomes.

Planetary systems have their own timescales. As the beginning of this chapter noted, carbon persists in the atmosphere over centuries. Similarly, biodiversity may take millennia to re-form once destroyed, and synthetic chemicals can persist in the environment for eons. As we strain the boundaries of various planetary systems, we create changes that can alter the entire planet for geologic spans.[31] In other words, the material facts we confront, the first element of a political problem (the first row in table 1.1), are shifting as humanity shapes the earth's fundamental systems for the first time in our existence.

Second, technological and scientific development have changed both our material ability to shape the future as well as our ability to measure and understand problems beyond the present. Technological changes like gene editing or AI, to name just some examples, have the potential to fundamentally alter human society. Once created and deployed, their effects may not be reversible.

The impact of human development on the planet or the advent of nuclear weapons are similar. New tools also allow us to alter the future (and indeed the present) much more profoundly, reshaping material facts.

In the same vein, science and technology also allow us to extend the timescale on which we perceive problems by enhancing both our understanding of the past and our forecasting abilities (the second element in table 1.1). To even begin to understand the risks posed by climate change we had to gain a deep understanding of the chemistry and physics of the atmosphere. We had to look back in time to understand previous phases of the earth's history through techniques like chemical analysis of ancient ice cores or air trapped in prehistoric rocks. We had to gather data from thousands of old handwritten weather observations written in dozens of languages and passed down through a range of oral traditions. And we had to build complex computer models to simulate what, based on our understanding of all of the above as well as the economy and society, might happen in the future. Only through this "vast machine" of human knowledge have we begun to grasp the danger we face.[32] As this example shows, changes in science and technology allow us to perceive more distant risks and impacts, lengthening the time span of how we understand problems.

Third, and more tentatively, social beliefs about how to value future generations may be shifting (the third row of table 1.1). We *perhaps* are starting to care more about the future. To be sure, attention to the needs of future generations is embedded in nearly all traditional human ethical systems. For example, scholars often point to the Iroquois maxim to consider the impact of a decision across seven generations. The general principle that we should care about our descendants is so common across belief systems that it can be considered a kind of basic moral intuition, like the injunction against wanton murder, that stems naturally from humans' reliance on social organization and perhaps even our biological imperatives. Strikingly, belief systems that disagree on many key points share an emphasis on the value of the long term. For example, modern conservatism and ecologism disagree sharply in countless ways, including about how to address climate change, but both agree that people in the present have a duty to consider how their choices will affect people living in the future. In the realm of normative philosophy, a groundswell of literature has emerged arguing that we should value the future more, not least because of the ways in which the Anthropocene and changes in technology have increased in the weight of the present on the future.[33] Indeed, a whole philosophical movement, long-termism, has emerged around this idea, connected to consequentialist beliefs that "future people count. There could be a lot of them. We can make their lives go better."[34]

Social scientists are less concerned with which particular belief system motivates an interest in the future and more concerned with how those beliefs do or do not shape behavior and institutions. Some prima facie evidence suggests these concerns may be growing. Today, 41 percent of written constitutions make reference to future generations, as well as hundreds of international legal instruments, a trend that sharply accelerated from the 1990s onward.[35] The UN secretary-general has proposed a Declaration on Future Generations. Recent surveys of both legal professionals and laypeople have found striking support for the idea that future generations should be protected in law.[36] Such changes in political beliefs and institutions can shape how we think about the length of problems. Because political problems are socially constructed, to the extent our norms and institutions place more value on future generations (the fourth row in table 1.1), we will treat more problems as long problems.

So if our problems are longer because they have changed materially, we have the technology both to shape the future and to understand distant impacts more accurately, and we may care more about the future and act on this belief politically, we must seek to govern across time. This book asks, can it be done? If so, how?

About This Book

This book brings the core insights of political science to bear on the problem of governing over time. Theoretically, it does not propose and test a single explanation but rather seeks to develop a general political economy account of long problems. Empirically, it does not examine a set of cases but rather draws on a wide range of illustrative examples from around the world, rooted mainly but not exclusively in the problem of climate change. In this way, it seeks to show why long problems are hard to govern and how we might nonetheless seek to understand their politics so as to advance solutions. These arguments aim to speak to scholars and analysts studying long problems, to policymakers grappling with them, and to students and citizens looking to understand them.

This focus connects to a long tradition of scholarship. Governing over time is a very old problem. But the modern idea that we can and should look ahead, and indeed seek to shape the future to our goals, grew out of the nineteenth-century scientific revolution and its promise that we could understand the world and, through human ingenuity and agency, forge some kind of "progress."[37] Later twentieth-century ideas on modernization rekindled this interest in understanding and shaping or even planning the future, just as reactions to some of the fruits of progress—for example, the threat of nuclear annihilation, environmental degradation—prompted calls to make forward-looking caution a fundamental principle for a society putting itself at risk.[38] Today, prompted in large part by the growing recognition of long problems, a wide range of disciplines are seeking to understand how human societies can best govern themselves over time.

The arguments in this book come from a political science perspective but one that takes seriously insights from a range of subfields and adjacent disciplines and one that seeks to speak to anyone interested in the problem of governance across time. Even though political science offers a powerful lens through which to study time problems, the field can also benefit enormously by looking at other ways of approaching the issue.

First, as discussed above, philosophers and normative political theorists have built a sophisticated literature examining why we should care about long problems and therefore raised the question of how they should be governed. Building from principles like the fundamental equality of human lives, or the duty not to interfere in the life chances of others, or traditional beliefs around stewardship, a surprisingly diverse array of philosophers have argued that present generations should care about future ones. By making clear that long problems *should* be governed and also by debating how we might best govern them, political philosophy has done much to put this topic on the agenda. Indeed, this book can be understood in part as an attempt to help the social science literature catch up with our normative colleagues.

Second, political scientists working in the realm of critical theory—a broad term for approaches that probe and question power, including nonpositivist approaches—have explored how time matters for conceptualizing power. For example, who wins and who loses from the instantaneous nature of modern financial transactions?[39] How are arguments around "prevention" mobilized to justify military interventions or policing?[40] How does the understanding of the future empower or disempower actors in the present? Following such questions, scholars speak of a "temporal turn"[41] in international relations theory.

Third, scholars working in the political science of the environment or in multidisciplinary environmental studies have grappled deeply with the temporal mismatch between human and environmental processes. This body of work has explored in detail the dynamics of prevention and of transition and has unpacked how uncertainty around future outcomes affects political decision-making. Scholars of climate politics have posited theories of overcoming lock-in or generating "sticky" solutions.[42] A growing body of work on "governing in the Anthropocene" tackles directly how political systems are changing—or not—to respond to problems like climate change that extend across both space and time.[43]

Fourth, and more generally, systems theorists—including Geoffrey Vickers, quoted above, who was a pioneer of the field—have explored many of the patterns that long problems raise. When considering systems overall, differential rates of change, as opposed to levels, are often the most important variables to study. Causal processes may exhibit positive or negative feedback effects. Tipping points may lead to fundamentally new equilibriums, creating nonlinear processes. All of these concepts help us probe the time-dependent assumptions that bind much social science scholarship.

While these four bodies of work are mostly in dialogue with each other, one contribution of the book is to gather their insights together and to translate them into the language of most social scientists, who stand to gain from taking them seriously. It is fair to say that each of them is significantly ahead of the bulk of contemporary political science, economics, or related disciplines in their conceptualization of time. As I argue in chapter 6, most contemporary social science literature makes fairly strong (implicit) assumptions about problem length that truncate the scope of our analysis and limit the generalizability of our theories across time.

That is a shame because there is in fact a huge amount that political science in particular can contribute to these questions. I aim to show how the discipline's central concerns—how and why actors develop certain interests, how they build, wield, and contest power to advance those interests, and how institutions structure their interactions—enrich our overall ability to understand long problems. In particular, this book focuses on understanding the political implications of long problems and analyzing how and under what conditions governance may emerge. Political science has much to give back to the wider intellectual community grappling with the dilemma of long problems. Studying the political economy of long-term governance can help fill a vital gap between normative ideals and abstract concepts on the one hand and the realities of how societies organize themselves on the other hand.

Although attention to long problems remains underdeveloped in the core of political science, there are significant exceptions. As chapter 6 discusses, the discipline possesses significant tools for studying time. In particular, this book takes as its point of departure three key works: Paul Pierson's *Politics in Time*, Alan Jacobs's *Governing for the Long Term*, and Jonathan Boston's *Governing for the Future*. Pierson not only provides a canonical treatment of path dependency but also sets out how to understand long-term processes of change and sequencing more generally. Jacobs analyzes the politics of intertemporal investments in democracies, creating both a theoretical structure and a rich empirical baseline for considering the politics of redistribution across time. And Boston encyclopedically surveys and evaluates mechanisms for how governance can be made less short term. I seek to build on these insights and other work in the field.[44]

The book proceeds as follows. Having presented the challenge of long problems and explained why they matter, in the next chapter I explore why long problems are hard to govern. It begins by compiling the various arguments made around short-termism in politics: why it exists and why it can lead to perverse outcomes. The bulk of the chapter, however, uses the concept of long problems to clarify three enormous political challenges. People in the future have only "shadow interests" in the present, limiting political agency that favors long-term outcomes. Dynamic problem structures that shift over time lead to institutions that lag behind what is functionally required of them. And because long problems require, by definition, action before their effects are felt, issues of uncertainty, low salience, and obstructionism are pervasive.

Chapters 3 through 5 then analyze strategies for governing long problems, corresponding respectively to the three challenges introduced in chapter 2: shadow interests, institutional lag, and the early action paradox. I start in chapter 3 with the last, as it encapsulates perhaps the core challenge of long problems: acting early. Making information about the future known and salient through informational tools or foresight processes can change action in the present when/if actors in the present have incentives to act in a long-term way, an important scope condition. But policymakers can also go further and use experimentalist governance techniques to confront the challenge of uncertainty directly, or deploy catalytic strategies and institutions, including those in the Paris Agreement, that can under certain conditions erode obstructionism by shifting incentives over time.

Chapter 4 turns to the challenge of shadow interests. Institutions that represent future interests in politics, either with reference to a specific issue like climate change or on behalf of future generations in general, can add an important element of agency to efforts to make knowledge of the future known and salient in politics. More powerfully, trustee institutions like courts and central banks can be given explicit mandates and powers to act on behalf of future interests. And a wide range of strategies can be used to actually extend political actors' time horizons, including forms of participatory deliberation like climate assemblies.

Chapter 5 focuses on the dilemma of navigating the tension between durability and adaptability to overcome institutional lag. Long-term goal-setting like the Sustainable Development Goals (SDGs) can drive continuity, while tools like sunset and review clauses can create opportunities for reflexive updating. Similarly, by incorporating automatic trigger mechanisms—such as indexing policy to certain trends or outcomes—into governance, policymakers can ensure there is an opportunity to update, while building up reserves, such as those we see emerging in sovereign wealth funds, can provide the capacity to do so.

Each of these three chapters surveys a range of governance solutions to a specific challenge long problems pose and examines the conditions under which and processes through which they can have more or less effect on the problem. I do not present a novel empirical examination of how we have in the past governed long problems but rather use the book's theoretical tools to examine how we might do so drawing on a range of examples. Throughout, my focus is not on abstract solutions but how and under what conditions specific tools may or may not reshape politics. As these chapters show, these tools are used by and available to all types of countries—democracies and autocracies, wealthy and developing countries—with a wide range of political cultures. As with all governance, however, state capacity is needed to deploy such tools effectively. These chapters represent the bulk of the book's contribution, drawing extensively on the example of climate change but also drawing in other issues for comparison.

Unlike the others, chapter 6 targets scholars and research students specifically, looking at which social science tools, both theories and methods, can already help us analyze long problems and how new approaches can deepen our understanding of them. It explains how taking long problems seriously both challenges current approaches and creates exciting opportunities for theoretical innovation. It emphasizes the importance of looking at rates of change, dynamic problem structures, and empirical study of future outcomes. Readers not seeking to analyze long problems themselves may wish to skim or skip this chapter, though social scientists will find a perhaps provocative challenge to expand our methodological and even epistemological repertoire.

Finally, chapter 7 concludes by considering what it would really mean to govern across time. Although there has been some progress in climate policy in the last decades, we are collectively falling woefully short. The policy ideas exist, but sufficient political support for them does not. The arguments in this book help explain why. The political institutions we have inherited are stacked against effective governance of long problems like climate change.

To really tackle climate or any other long problem, we need to change the rules of the game.

I therefore propose an institutional agenda on climate change to help build the political conditions under which effective policy becomes more feasible. This agenda builds on the tools examined in chapters 3–5: weaving together future-oriented informational systems across the policymaking process, experimental and catalytic strategies and institutions, ways to represent future generations and create trustees for them that have real power, processes that can extend time horizons such as participatory deliberation, frameworks to set and continuously update long-term goals and pathways toward them, triggers to keep us on course, and new reserves to enable investment in transition and resilience. Together and over time, this family of long-term governance reforms could remake our political institutions in profound ways, reaching beyond a single issue like climate change and reorienting politics overall toward long-term interests.

However, the threat of climate change or any other long problem will not necessarily drive us toward this governance transformation. Indeed, we have reason to expect that as climate impacts and decarbonization grow more intense and existential, political pressure for immediate reaction and protection will make our political system more short-termist, not less. Instead, what the climate challenge does present is an opportunity for policymakers and citizens to catalyze more long-term governance systems going forward. The choice to do so or not is ours.

The book ends by considering the possibility that we succeed. Throughout human history, profound changes in political "technology"—the nation-state, representative democracy, global governance—have tended to lag changes in economic and social systems. But if we take governance of time seriously, then political decisions and activities can increasingly shape the social and economic future. Although there is always mutual feedback between these macrosystems, improving society's capacity to shape material and social outcomes in the future—that is, to govern time—can create unprecedented possibilities, perhaps both good and bad, for our collective agency.

2
Why Long Problems Are Hard to Govern

Starting in 1990, the Intergovernmental Panel on Climate Change (IPCC) has issued six assessment reports summarizing the state of climate science and what can be done to mitigate the problem and adapt to its impacts. These reports, which cover the entire scientific literature, have repeated essentially the same message in 1990 (1st), in 1995 (2nd), in 2001 (3rd), in 2007 (4th), in 2014 (5th), and in 2023 (6th), albeit with increasing precision. We are causing climate change. It will be devastating. We can prevent it by reducing our GHG emissions. We can reduce our suffering if we adapt to its impacts. We can be confident the seventh assessment report, due in 2028, will agree.

Despite decades of warnings, annual emissions have nearly doubled globally since 1990, pumping a quantity of GHGs into the atmosphere roughly equivalent to all the emissions from the Industrial Revolution to the end of the Cold War. Why have we not done more?

This chapter explores the political challenges that long problems pose. It is easy to see that long problems are hard to govern but harder to say exactly why. We intuitively understand the problem as excessive short-termism, both in our own lives and in society overall. But this begs the question, why do we individually and collectively suffer from a "presentist bias"?[1]

There are many drivers behind short-termism, and the chapter first reviews the dominant ideas in the literature. But understanding why long problems are hard to govern—that is, why it is hard to create collective rules and interventions that steer outcomes to our long-term interests—requires more specific analysis. Although short-termism has all too many causes, the chapter seeks in particular to develop a political economy analysis of long-term governance.

This lens focuses our attention on what shapes political actors' preferences, how they organize to pursue their interests against those of other actors, and how institutions mediate this contestation. I organize the argument around three core political dilemmas of long problems. First, governance is often driven by people and groups strongly affected by a problem, but people in the future lack any agency over decision-making today. They only have *shadow interests* in the present. Second, long problems shift over time, but the institutions set up to govern them can be hard to update. This combination of dynamic problem structures and "sticky" institutions creates an *institutional lag* between our governance tools and the challenges we face. Finally and perhaps most significantly, long problems suffer an *early action paradox*: by definition, they require action before effects are felt, but this means we must act under conditions of uncertainty, low salience, and obstructionism. Chapter 3–5 return to the three dilemmas this chapter raises—starting with the early action paradox, as it is perhaps the most central—considering how we can face them and under what conditions we can prevail.

Short-Termism: The Tyranny of the Present

Policymakers, social scientists, psychologists, and everyday observation all tell us that as individuals and as societies, we often struggle to look beyond the immediate. Famously, studies show that most children cannot resist a marshmallow placed before them for more than a few minutes, even if they know that waiting just a few more minutes would earn them an extra marshmallow. Market analysts worry that CEOs often seek to maximize quarterly earnings reports instead of longer term value. And many observers of politics point to politicians focused more on tomorrow's headlines or next year's election than on the long-term good. Scholars have lamented that society is in a state of "time denial" in which too many of us are "time illiterate," having little sense of human history, much less planetary history.[2]

A long problem like climate change seems to demonstrate perfectly the dangers of short-termism. Because of the temporal gap between cause and effect, reducing emissions now prevents things from getting worse in the future. This lag means that actors in the present will have to wait to see the benefits of any contributions they make to preventing climate change. Adaptation to climate change is similar. Many investments to reduce climate impacts, like a new seawall or a restored coastal marsh that can absorb floods, could take years or decades to yield benefits, which only accrete over time. For example, in 1953, a devastating flood from a storm on the North Sea swept up the river Thames, causing nearly two thousand deaths across Europe and the equivalent of billions of dollars of damage in today's terms. In response, the UK government studied the problem and ultimately decided to make a barrier across the river in East London, a project not finished until 1982.[3] Although three decades late for those killed or displaced by the 1953 storm, the barrier has proven to be an increasingly valuable investment, being deployed more and more each decade.

Short-termism can be particularly vexing when it reinforces other aspects of problem structure. In climate, for example, short-termism reinforces the distributional challenges of acting because there is a significant power imbalance between those who cause the problem and those who bear its brunt. Both within countries and across them, to a first approximation, emissions are correlated with economic output and consumption and therefore with political power. Meanwhile, those countries and communities most vulnerable to climate change are those with least capacity to adapt, which is in turn correlated with a lack of economic weight and political power. Moreover, many of the future benefits of climate change are not tangible gains, like an artist's magnum opus that emerges after years of hard labor, but rather take the form of a lower level of risk compared to a modeled counterfactual scenario. So climate policy is not only a struggle to get actors in the present to do things that will yield benefits only in the future; it is a struggle to get the most powerful actors in the present to do concrete things that will provide intangible benefits to weaker actors in the future.

For these reasons, temporal mismatches are just as much at the heart of the climate challenge as the dilemma of collective action, and short-termism is just as much a problem as free riding, though the latter is more frequently discussed. As Mark Carney, the former Bank of England governor turned UN climate envoy, has noted, the issue is not just a "tragedy of the commons" but also a "tragedy of the horizon."

In *Governing for the Future*, Jonathan Boston comprehensively shows how this short-termism is overdetermined in our societies, embedded in both human psychology and institutions. Before turning to the political economy of long problems, it is useful to review these factors.

Consider our minds. Most social scientists understand the world with some implicit or explicit degree of discount rate, meaning we tend to think that actors value a thing in the present more than they value that same thing in the future. I return to the analytic uses and abuses of discount rates in chapter 6, but for now, it suffices to note the range of reasons why it may be true that we value proximate things over more distant things. Most simply, people may have so-called pure time preferences, caring more about present welfare than future welfare. In other words, we may just really want that marshmallow *now*, not five minutes hence. The empirical evidence on this idea is mixed, however, as people's time preferences seem to vary across different areas of their lives in inconsistent ways. Children want marshmallows now, but people can also save money over years to support their children's education, for example. Preferences also seem to be hyperbolic, meaning they drop off sharply at first but then essentially remain constant long into the future. So while there is some evidence for "pure" short-termism at the individual level, it seems to operate in complex ways.[4]

Behavioral science helps explain some of these oddities. Bounded rationality leads us to expect our minds to deviate from rationally distributing values over time and instead to rely on numerous heuristics and cognitive shortcuts. Unfortunately, many of these also have the effect of overprioritizing the present. For example, people anchor their preferences on the status quo and are averse to losses, so they have a natural bias toward the present state of affairs even if it is suboptimal. People are also bad at accurately making the kinds of probabilistic judgments needed to weigh future options, particularly relative risks. And with limited cognitive bandwidth, we focus our scarce attention on the immediate over the distant.[5] As Boston notes, quoting Al Gore, "the future whispers while the present shouts."[6]

Finally, even when we can perceive, understand, and value a long-term interest, it does not necessarily follow that we are able to act in a way to achieve it. Most of us know we would be better off exercising more, eating better, saving more money, and preparing tasks more in advance, but we still struggle to optimize these ideas in our daily lives. We often use commitment devices to try to force ourselves to act more consistently with our longer term goals, but the pull of the immediate is always there.

Although the limits of the human mind are clearly relevant when considering long problems, the short-termism of human institutions—our businesses, organizations, governments, and so forth—are even more significant. After all, it is on the organizational and societal scale that long problems will or will not be solved. However, the challenges these entities face are not just larger manifestations of the individual-level short-termist mechanisms identified above, though there are some analogues. Instead, institutions and organizations have their own features that create presentist bias.

Perhaps the most important one is the way many institutions are organized around relatively short-term evaluation and accountability cycles. Publicly traded companies must make the case for ongoing investment to their shareholders each quarter. Schools are assessed regularly by government accreditation bodies, often via standardized tests. Nonprofit groups report to their board. Government departments must justify their performance and budget to leaders each year or two. And most political leaders and parties need periodic renewal every few years, either through elections or, for less democratic governments, the formal or informal validation of interest groups and elites.

On this latter point, there is a stubbornly persistent but unpersuasive "democratic myopia" thesis that elections in particular breed short-termism and that authoritarianism may therefore offer some insulation from its dangers.[7] Although it is certainly true that elections can create short-term incentives, there is very little evidence that an absence of elections can avoid them.[8] Just like democratic politicians, authoritarian governments require support to endure. Support may come from economic elites, from the military, from different segments of the population, or from foreign powers. The cost of support is some degree of accountability (albeit perhaps to a very narrow core of supporters), which in turn requires periodic evaluation. Should key supporters grow discontent, authoritarian rulers must either seek to repress them (a costly and potentially risky strategy) or respond to their needs. In this way, short-term pressures rise onto the autocratic agenda just as surely as they appear in electoral politics, though often with less predictability and regularity. Moreover, even if authoritarians could insulate themselves from short-term pressure, it does not necessarily follow that they would then have incentives to promote long-term welfare. Indeed, most studies show the authoritarian regimes tend to provide fewer public goods than their democratic counterparts.[9]

Looking across the whole evidence base suggests that there is much more variation in short-termism within both democracies and autocracies than between them. Researchers correlated an index of intergenerational solidarity with democracy and found, if anything, a roughly positive correlation between democracy and policies like investment in education, pensions, or environmental protection.[10] As we might expect, factors like a country's level of development and the quality of its bureaucracy roughly correlate with long-term governance as much as regime type.

The supposed autocratic advantage in long-termism is typically rooted more in a (largely valid) critique of electoral politics than a rigorous assessment of how autocracies really work in practice. Indeed, a more productive avenue of analysis is to look in detail at how decision-making in a political regime, democratic or autocratic, is structured. For example, some research shows that proportional electoral rules give politicians greater electoral safety and so makes them more willing to impose short-term costs to secure longer term gains.[11] Similarly, countries with more institutionalized political parties invest more in technology development—which yields benefits only in the long term—than countries where parties come and go. The logic is that generations of politicians overlap in institutionalized parties, so those about to step down share a political interest with the newcomers eager to take their place. This creates a greater incentive to support technological development—which costs money in the present but provides benefits only in the future—than in countries where parties do not endure.[12]

But what about China or Singapore? Both countries are frequently cited as examples where political leaders, insulated from short-term pressures stemming from competitive elections, can take a longer view. Unlike most autocracies, these governments do indeed prioritize long-term planning. But even in this context, the constraints of relatively short evaluation cycles are widely observed. The Chinese government puts a five-year planning cycle at the heart of its policy process, issuing regular five-year plans that outline key policy objectives and targets against which officials will be judged. Although this system has some merits, it also suffers from short-term pressures. Because officials are evaluated on how successfully they meet the plan's goals, they tend to rush to complete their targets as the five-year deadline approaches. For example, at the end of 2020, as the deadline for meeting the targets of the thirteenth five-year plan approached, a number of local governments simply shut down electricity production to meet energy efficiency targets, resulting in the worst blackouts China has experienced in decades.[13]

As this example highlights, the real difficulty is that some kind of regular reporting and accountability is required for complex organizations such as governments to operate effectively. If shareholders could not count on regular performance reports from management, they would be loath to invest in a company. If regulators cannot regularly audit a school's performance, how can parents trust the teachers to educate their children? If political leaders have no way to manage bureaucrats, how can they achieve their political goals? And if constituents have no means for holding leaders to account, then political outcomes will be suboptimal. The principal-agent dynamics that pervade nearly all human organizations of any size or complexity force us to create institutions that build incentives around monitoring performance in an ongoing fashion. In other words, relatively short-term evaluation and accountability cycles are often necessary for institutions to operate effectively, even if they have perverse side effects. Short-termism may be a bug, but it is also a feature of a complex human system in which accountability relationships are required.

Other dynamics push human institutions toward short-termism as well. Analogous to individual attentional scarcity, organizations can process only so many issues at once. After all, they have only so many staff to execute operations, only so many meetings and processes to come to collective decisions, only so many days in a legislative session, and only so many hours in a day for leaders to make decisions.[14] Students of bureaucratic politics show how the timing of most decisions is not driven by their substantive importance so much as when deadlines, forcing moments, or windows of opportunity arise to require a decision, biasing attention toward the urgent.[15] On top of this, the media raise the political salience of only a few issues at a time, forcing certain issues to the top of decision-makers' priorities and leaving others languishing in the shadows. Indeed, the organizational or political "attention deficit" is even more stark than it is for individuals since the number of issues is greater and the steps required to assess and implement a course of action longer and more complex. This attentional scarcity is part of why agenda-setting is so important for political outcomes. Unfortunately, the less attention there is to spare, the more urgent issues will be focused on at the expense of less urgent issues, regardless of their relative importance.

Finally, like individuals, institutions struggle to impose longer term constraints on themselves because of the temporal inconsistency of preferences. A CEO may wish to invest in long-term growth but find banks will not lend to her without sufficient short-term cash flow. A politician may be elected promising fiscal rectitude but then face a close election in which short-term job growth could be the deciding factor. A new government in a postconflict context may sincerely wish to uphold human rights but then face overwhelming pressure to crack down on opposition groups as its hold on power becomes tenuous. Just as individual preferences are inconsistent over time, so are the incentives that shape institutional decision-making.[16]

Governments therefore seek numerous solutions that allow them to make credible commitments, mostly by accepting limits on their freedom of action in the form of institutions like independent central banks or judiciaries, or international institutions like human rights tribunals, investment treaties, or trade bodies. But these tools are often imperfect and prone to reversal because governments find it very hard to credibly limit their own power. Sovereign in their borders, states' inability to limit themselves leaves them at the mercy of their own short-term whims.

In sum, short-termism is pervasive at both the individual and institutional levels. Even worse, these myopic timescales can reinforce each other. For example, scholars have found that where people trust the state to deliver long-term benefits, they are happy to support long-term policies. But where there is political uncertainty over whether political leaders can and will provide long-term benefits (e.g., not raid a pension fund for short-term projects), individuals' policy preferences show a parallel presentist bias.[17] Or consider the role of the media as a vector for political communication and preference formation. A politician may wish to pursue long-term goals but also knows she will be either praised or vilified in the press based on what she does in the moment. With opponents launching attacks on social media or cable news instantly, her incentives will press against long-term approaches. In these ways, present-biased dynamics at the individual and institutional levels can reinforce each other, creating a broader "lock in" effect[18] that makes it very hard for any one actor or institution to get beyond short-termism. After all, if you live in a society that struggles to commit itself to the long term, what value can you place on a long-term outcome? You too must prioritize the short term. For a world facing a rising number of long problems, this collective short-termism is a very dangerous state of affairs.

It is important not to overstate the inevitability of this danger, however. To understand how to overcome short-termism, we also need to understand its limits. Even though short-termism is pervasive, evidence shows it is also neither omnipresent nor monolithic.[19] These exceptions are important because they can teach us about ways through and beyond short-termism. At the individual level, humans may have "marshmallow brains," as the philosopher Roman Krznaric puts it, but we also have "acorn brains"—a capacity for long-term planning that no other species possesses.[20] Studies find that our brains spend many "idle" hours thinking through possibilities and potentialities—for example, late-night worrying—in a way unique to humans.[21] Human institutions are also capable of overcoming short-termism. Despite all the challenges, we do sometimes carry out long-term plans. Building on the success of the Thames Barrier's success in its first forty years, the UK government is now planning to extend and augment it so it will remain effective through the end of the century. Throughout history, humans have built cathedrals, pyramids, canals, and other great works even though many who toiled to make them never saw the end result. Certain religious groups, universities, and other human institutions are designed with a generation-spanning view. Chapters 3–5 draw on some of these exceptions as they propose strategies for governing long problems. Social scientists and policymakers should not accept short-termism as a given but rather identify its drivers and determinants to devise strategies for meliorating it.

To do so, we need a more specific analysis of the political impediments long problems create for effective governance. These barriers are not just a failure of leaders or societies to take a long-term view. Instead, they are specific political mechanisms rooted in the logic of how governance does—and does not—happen: shadow interests, institutional lag, and the early action paradox.

Shadow Interests: Decision-Takers Have No Power over Decision-Makers

At heart, politics is the struggle of different interests to amass and wield power. Long problems are perhaps unique in that one set of interests, those most affected by a problem, are by definition absent from this process. They are not merely disenfranchised or marginalized; they simply do not yet exist.[22]

I term these interests "shadow interests" for two reasons. First, like our shadows, they have no weight or agency in decision-making. They can only follow where we go. Second, like the shapes flickering across the wall of Plato's cave, shadow interests are only dimly, subjectively perceived. Their exact form is unknowable, so different people can see them in different ways. In other words, even if we believe future people matter, we can only perceive their interests through our own projections. This one-sided nature of long problems makes them more difficult to resolve because the set of actors who would usually be a key part of driving toward a solution do not just lack political power, but they lack any agency whatsoever.

A 2013 report by the UN secretary-general recognized this conundrum:

> Future generations are politically powerless, with the representation of their interests limited to the vicarious concern of present generations. As stated in the report of the World Commission on Environment and Development, entitled "Our common future," "we act as we do because we can get away with it: future generations do not vote, they have no political or financial power; they cannot challenge our decisions" (A/42/427, annex, para. 25).[23]

If we consider some of the most important "non-long" problems human societies have faced, the significance of shadow interests in long problems becomes clear. For example, the nineteenth-century struggle to abolish slavery in European colonies and in the Americas shows how even deeply oppressed but existent interests were critical for change. Even though enslaved people of course lacked access to political power, they retained agency, and that agency played a key role in ending the institution of slavery. Most directly, figures like Harriet Jacobs or Frederick Douglass in the United States, Olaudah Equiano and Mary Prince in the United Kingdom, or José do Patrocínio in Brazil escaped bondage (or in the last case descended from those who did) and became advocates for abolition, influencing powerholders within the political system.[24] At the same time, people regularly revolted against their enslavers, seeking to free themselves by force. The fact that such efforts typically failed did not mean they were unimportant. While historians continue to debate the direct effect of such revolts,[25] the mere potential for enslaved people to exercise their agency through revolution meant that slaveholding societies had to invest vast resources in violent systems of repression. The difficulty of maintaining such systems, both economically and morally, hastened their demise. An enslaved person might have lacked political power of any kind, but his or her mere existence and (potential) agency were a critical and tangible part of the problem structure around ending slavery. As such, they were qualitatively different from the shadow interests of people who do not yet exist.

Numerous other examples demonstrate how the agency of affected people is a core part of how many political challenges get solved. For example, it is difficult to imagine the emergence of the modern welfare state in the twentieth century had marginalized populations bearing the brunt of inequality not organized themselves into political parties and won power. Nor would we expect pollution controls or health standards to be as strong had the victims of, say, chemical spills or tainted food not come forward to demand justice. In most countries, regulations now give those affected by such problems the ability to bring perpetrators to court.

Even when those affected lack direct agency, the fact that others can see and empathize with their suffering—as we do in countless humanitarian disasters, human rights violations, or crimes against humanity—creates a nontrivial tool to drive political action.[26] Under conditions of interdependence, decisions made in one part of the world can have sharp consequences for people far away who may have very few levers to affect especially the decisions of powerful countries.[27] For example, subsidies to protect farmers in rich countries can undercut the livelihoods of subsistence farmers across the world, or intellectual property rules to protect pharmaceutical companies can deprive whole populations of vital medicine. And yet these people nonetheless possess agency to resist, to move elsewhere in search of better conditions, or to seek to shame the powerful into recognizing their needs or otherwise intervening.

Imagine how each of those examples might have been different had the people affected not yet existed. Shadow interests are a problem because they cannot work via the usual mechanisms through which political contestation occurs. In Nobel Prize laureate Ronald Coase's famous hypothetical world of no transaction costs or power hierarchies, problems are solved efficiently because those negatively affected by a decision or action can seek compensation from those who benefit from it. In this way, all "externalities" can be efficiently "internalized" by the (assumed) ability of all parties to effortlessly make Pareto-enhancing contracts.[28] But this logic breaks down when those negatively affected simply do not yet exist. Who will sign the contract?

Coase's thought experiment perfectly shows how long problems thwart many conventional models of politics. For example, shadow interests help explain why we have not yet been able to resolve climate change through bargaining and are unlikely to do so. If we accept the conventional framing that mitigation creates costs in the present but avoids much greater costs in the future, a bargaining equilibrium should, in theory, exist in which those more affected by climate change compensate those causing it in order to get them to stop.[29] This kind of solution is often highlighted by, for example, students of trade theory. Trade liberalization benefits internationally competitive businesses but hurts those that are less prepared to deal with imports. But if the total gains outweigh the total costs, the industries that benefit should be able to compensate those that lose out in such a way that at least no one is worse off and overall welfare is higher. Even in the trade realm, however, such bargains are rare in practice.[30]

In climate change, they are practically impossible for several reasons. As noted above, the future benefits for any given actor are uncertain (reduced risk compared to modeled counterfactuals), while the present costs are highly concrete. Those who would be negatively affected by action now have concentrated, powerful interests, while those who stand to lose out from climate change are weaker and more dispersed. But perhaps most centrally, the affected interests who would need to pay off polluters in the present simply do not yet exist. A "nonsimultaneous exchange" is therefore required.[31] Such exchanges are theoretically possible, but the longer the problem, the more sustained they must be. For multigenerational exchanges, perhaps the only hope of approximating a bargaining solution is a kind of diffuse reciprocity in which the current generation, perhaps motivated by respect and gratitude for the contributions previous generations have made to our welfare, "pays it forward" by making similar investments for the next generation. Such a dynamic is certainly not alien to human nature but is difficult to achieve politically and likely depends on strong normative beliefs that privilege the welfare of future generations.

Of course, bargaining is not the only way to achieve political change. As the above example of abolishing slavery attests, activism, moral suasion, revolution, and war are just some of the pathways through which political contestation can come to a resolution. But none of these tools are available to shadow interests. Philosophers have accused the present of colonizing the future just as imperial powers colonized large swaths of the world, ignoring the presence of the people living there just as we ignore the future existence of people on the receiving end of long problems. Although the analogy may hold on the ethical dimension, the political dynamics are quite different. Colonized people eventually won their independence. But future people have no more hope of persuading us to make better choices for them or taking up arms against us to force us to do so than we may hope to alter the choices of previous generations. As Roman Krznaric writes, people in the future "cannot throw themselves in front of the king's horse like a suffragette, block an Alabama bridge like a civil rights activist or go on a Salt March to defy their colonial oppressors like Mahatma Gandhi."[32]

In sum, shadow interests are not the same as weak interests. They are qualitatively different because they lack even latent agency. Because they do not yet exist, they cannot negotiate or militate for their goals. No one needs to repress them to ensure they do not get their way. To the extent the harmful impacts of long problem like climate change fall on shadow interests, many of the usual tools of politics will be unavailable.

Institutional Lag: Dynamic Problem Structures

We typically analyze political challenges by defining their "structure" in terms of characteristics like the set of relevant actors, the beliefs and information those actors possess, the distribution of their preferences and power, the institutional and material environment they inhabit, and as I argue, the length of time between causes and effects. As problems lengthen, however, none of these other elements can be taken as constant. Social scientists and policymakers often assume, implicitly, that a problem's elements are fixed to make them more tractable for analysis or strategy. But as problems lengthen, that assumption grows less tenable. At the extreme, all "things" are, on some timescale, "processes." As one philosopher of time notes, "the names we give to processes—the Himalayan Mountains, a kurrajong tree, a ladybug—stand for families of parameters stable for shorter or longer periods of time."[33] Indeed, at the most fundamental level, the universe itself is changing and expanding. The physical conditions that governed its first moments are not ones that affect us today. This is of course true as well for the various elements of a political problem, especially human and organizational phenomena like beliefs, interests, and power relationships. We often operate at timescales that make these factors relatively constant, but over the long term, they all change sooner or later.

Shifting problem structure creates one specific challenge for governing long problems: it increases the mismatch between the structure of a problem and the institutions we create to address that problem. Addressing political problems involves rules and organizations that structure actors' norms, expectations, and incentives, as well as the information they possess in regular ways. Such institutions define the "rules of the game." Constitutions determine who holds power and how other rules are made. Laws and regulations govern individual and economic behavior, potentially backed by the coercive power of the state. Treaties define mutual obligations between nation-states, while international organizations allow states to jointly address shared concerns. Nation-states are hardly the only actors creating such rules and organizations. Cities, businesses, civil society groups, and other sub- and nonstate actors institutionalize themselves in various ways, both within a country and through transnational networks and initiatives. Critically, institutions need not be written on paper or otherwise formalized. Informal institutions play a critical role in structuring politics as well.

For example, a pandemic may require international information-sharing agreements, rules restricting physical interaction, requirements on testing for disease, pooled programs to ensure vaccines reach the poor, and so forth. Climate change, of course, is implicated in a vast "regime complex"[34] (or "transnational regime complex"[35]) made up of countless partially overlapping institutions, including national laws and policies,[36] bilateral, plurilateral, and multilateral treaties, sub- and nonstate rules and organizations, transnational networks and initiatives (many involving multiple types of actors),[37] and intergovernmental organizations, among others. In this way, climate change is typical of a number of "global issues" that have seen a proliferation of cross-border institutions over the last decades.[38]

Institutions are therefore both a part of what makes up a political problem's structure (recall the fourth element defined in chapter 1) as well as essential tools political actors manipulate to address them. Political scientists seek to understand how actors seek to create and shape institutions to suit their goals, explaining institutional outcomes through a mix of what they do (functions) and what their creators want them to do (interests). Functionalist theories expect institutions to "match" the problems they seek to address. For example, in international relations theory, students of rational design identify regular patterns between institutional characteristics and problem structure (e.g., more flexible commitments when uncertainty is higher).[39] After all, functionalists reason, actors would only bother to create and maintain an institution if they thought it could help them advance their goals.

Institutions also reflect power relationships. Because the rules of the game can have sharp distributional consequences, powerful actors seek to ensure they serve their interests. But even though this means that institutions tend to reflect the interests of the powerful, they can also put limits on power by making its exercise more regular, predictable, and legitimate. Therefore, for reasons of both efficiency and power, we should expect institutions to reflect other elements of a problem's structure, including its functional needs and the constellations of power and interests around it.

On the other hand, institutions, once created, also help to define a problem's structure and can be sources of power. The UN Security Council, created in the aftermath of World War II, was explicitly designed to reflect the interests of the dominant powers. The victorious powers gave themselves privileged seats. Those positions then became sources of power in later decades, as even countries that lost their superpower status maintained a veto over global security matters that came before the council. The United Nations Framework Convention on Climate Change (UNFCCC), in contrast, never adopted formal decision-making rules, instead defaulting to a norm of consensus decision-making that gives every country, in effect, a veto.

As these examples highlight, institutions are meaningful elements of a problem structure in part because they are sticky. Indeed, some degree of durability is assumed in the very concept of an institution and is both a core part of why institutions play such a central role in politics and also an inevitable product of the way institutions are created and changed by political actors.[40] To be able to influence actors' information, beliefs, and incentives, they cannot be merely transient reflections of a momentary constellation of power and interests. If you knew a law was here today but gone tomorrow, it would not seriously affect your planning. Similarly, if a political party expected a certain constitutional arrangement to fail, they would not build their strategy for winning power around it. All human institutions have some kind of half-life, though these vary significantly. On average, laws may change on a scale of years, national constitutions or international organizations on a scale of decades, and states on a scale of centuries. Informal norms may last for centuries and then fundamentally shift within a generation—for example, around slavery, women's role in society, racial equality, or sexuality. Persistence over time is a key part of why institutions are meaningful.

Institutions are sticky also because, once created, they are typically time consuming and difficult to change.[41] National constitutions by design tend to require special procedures and supermajorities to amend. But even basic laws and regulations can only change with a mobilization of political interests, contestation or bargains across groups, and the nontrivial procedural steps of drafting, legislating, or rule-making, and potentially, legal challenge and review. At the international level, institutional change is often even more difficult because it tends to require agreement across heterogenous national interests. The UN Security Council, for example, has resisted reform since its creation despite many attempts to modify its membership or the use of the veto. This kind of institutional inertia is a key driver of gridlock in global governance generally.[42]

Studying how and why institutions change or not is a major concern of social scientists. In contrast to purely functionalist or power-based approaches, historical institutionalist frameworks emphasize not how institutions reflect a certain functional need or power structure but rather how they accrete and evolve over time.[43] Instead of smoothly "updating" as needs or interests evolve or as they are displaced by superior competitors, institutions can follow their own partially endogenous logic. For example, choices made at one moment can alter the choices available at later moments. Instead of optimizing across all possible equilibria at every stage, actors may find certain options more or less available depending on previous choices. This "path dependence"[44] is not necessarily determinative but can significantly shift institutional outcomes away from what a purely functionalist or power-based account would imply. Again, the Security Council provides a case in point. By institutionalizing the privileges of the victors of World War II, the council made itself very difficult to alter (largely by design), even as geopolitics changed radically. So, while institutions can change, they often resist it. And when they do change, they do not necessarily align with the underlying functional needs and power structures around them but rather reflect their own path dependencies.

This logic means that over time, institutions have scope to become increasingly disassociated from other elements of a political problem. Often, scholars conceptualize this dynamic as a form of punctuated equilibrium.[45] Institutions match the functional needs and power relationships that exist at a certain point in time, but as time goes on, functional needs and power relationships shift, making the institutions less and less fit for this purpose. Eventually, this gap contributes to some kind of crisis that "punctures" the status quo and from which a new institutional equilibrium emerges. For example, students of power transitions in international relations note how international orders reflect the power and interests of the dominant state or states but then, as the balance of power shifts, often descended into crises from which new orders emerge.[46]

This dynamic interacts with problem length. If institutions have divergent effects in the short term and the long term and the institution's creators have short time horizons, then the longer-term effects of institutions may be mere by-products.[47] Moreover, the longer the time between a problem's causes and effects, the more chance there is for interests and power to shift as, for example, technology evolves, norms change, economic models wax or wane, or, indeed, as our climate changes. The exact size of the gap of course depends on the relative rate of change between these other factors that make up a problem and the rate of institutional change. But anytime the former is greater than the latter, increased problem length will be associated with a greater institutional lag. To the extent institutions are sticky, they will not rapidly "self-correct" to respond to changed functional needs or power structures, and moreover, any changes that do occur can be influenced as much by previous institutional arrangements as current requirements. The longer the problem, the greater scope for such mismatches to grow. In this way, the natural stickiness of institutions—critical to their effectiveness—has a perverse interaction with problem length.[48]

Multilateral governance of climate change provides an excellent example of institutional lag. Created in 1992, the UNFCCC already showed a certain degree of "inherited" path dependence at its birth. The mental model guiding its initial design was inspired in the ozone regime, created just five years earlier, and widely seen as successful case of multilateral environmental governance. Countries believed this "convention and protocol" model, in which a broad framework convention was followed by a series of increasingly specific negotiated deals that locked in firm commitments, was the best way to overcome what they perceived to be a collective action problem. The history of previous institutional choices thus powerfully informed countries' understanding of the problem and their preferred solution.

The new institution's governance quickly became a flashpoint of contention, with developed and developing countries failing to agree on formal rules of procedure. Overall, countries were wary about binding themselves to decision-making processes they could not control. The UNFCCC therefore defaulted to a norm of consensus-based decision-making, which basically means that nothing is agreed until every country agrees on every point. This decision rule makes the UNFCCC very sticky indeed.

This stickiness has produced a number of gaps between the needs of the problem and multilateral arrangements. Perhaps most obviously, the early

UNFCCC created a divide between the obligations of so-called Annex I countries (forty-three industrialized states) and all others, with the former required to move first on reducing emissions. This arrangement, again inspired by the ozone regime and previous environmental treaties, seemed to make good sense at the time. In 1992, the Annex I countries made up 54 percent of global emissions. In 1997, when the Kyoto Protocol was adopted, this was still 45 percent. In 2009, when the Copenhagen summit ended in stalemate after the failure of developed and developing countries to agree on a differentiation of responsibility, Annex I countries accounted for only 36 percent. At Paris, it was 32 percent. Whereas the general principle that industrialized countries should move first makes good sense (given that they have more capacity to do so and have benefited disproportionately from carbon-intensive development), which countries are and are not industrialized has changed significantly since 1992, while the institutional rules have been slow to adapt.

Another example, noted above, is the difficulty of shifting the definition of the climate problem. The initial focus of the UNFCCC was essentially entirely on reducing emissions. Vulnerable countries had to fight to include consideration of climate impacts in the process. It was not until the adoption of the Paris Agreement in 2015 that a UNFCCC outcome reflected relative parity between mitigation and adaptation. At present, the inclusion of compensation for loss and damage from climate change is fiercely contested. Other issues, like geoengineering, remain taboo. These issues are increasingly discussed in science and policy around the world but, given the stickiness of the UNFCCC, will only work their way slowly, if ever, into the multilateral discussions.

Going forward, the elements of the climate problem will continue to shift. I explore this further in chapter 6. We can say with certainty that climate impacts will be significantly worse in thirty years, but the array of material and social facts, problem understandings, and political dynamics is highly uncertain. As those factors continue to change, the stickiness of our institutions will create a growing lag between institutions and other elements of the problem.

The Early Action Paradox: Uncertainty, Salience, Obstructionism, and Muddling Through

The final governance challenge long problems pose, and perhaps the most important, derives from the difficulty of mobilizing political action before a problem's effects are known and felt. Because long problems stretch far into an unknowable future, it is more difficult to act decisively on them now. After all, how can we know which course of action is best? And how we can mobilize support to change things when the effects are so far off and so many more urgent priorities demand attention? Uncertainty and low salience in turn enable obstructionism, giving interests opposed to a certain course of action favorable political conditions to resist change.

But at the same time, long problems require early action. If causes and effects are separated by long periods, only early action can meaningfully influence outcomes. Waiting until effects manifest is, by definition, too late. Long problems therefore pose a governance paradox. Their greater uncertainty and muted urgency limits agency in the present and bolsters obstructionism, but early action is precisely what is needed to meaningfully affect outcomes given the long lag between cause and effect.

Consider first the relationship between problem length and uncertainty. All else equal, longer temporal distance between a problem's causes and effects leads to greater uncertainty because the elements that make up a problem can shift either endogenously or exogenously, as the previous section discussed. Technologies emerge and diffuse. Beliefs change. Power transitions from one set of actors or countries to another. Some of these changes may be quite specific to the problem at hand; others may stem from broad shifts in the "state of the world," as the rational design literature or common game theoretic approaches term it.[49] Long-termism is therefore a significant epistemic challenge.[50]

This uncertainty limits the ability of actors to shape their environment through purposeful action. As the temporal range of the effects of political choices increase, we need more time to consider their effects, particularly if they might be irreversible.[51] Actors cannot be sure that their efforts will have the desired outcome, and this doubt means they may be less willing to act, especially when action is costly. Instead, they "muddle through."

As political scientist Charles Lindblom's famous 1959 article with this title argued, even for non-long problems, policymakers rarely make every decision following a hyperrational, comprehensive, utility-maximizing policy development process.[52] Instead, they consider a few proximate examples and choices and make an incremental step toward what they judge best. For long problems characterized by significant uncertainty, the hyperrational model Lindblom critiqued as unrealistic is additionally inappropriate. Instead, we can expect a number of sequential, incremental, boundedly rational choices to be the norm.

But the longer the problem, the worse this muddling through approach, or what Lindblom called making "successive limited comparisons," will fare because only early actions can substantially shape the outcome. Urban planning demonstrates this challenge well. Deciding what to build where has long-lasting implications for city life, and all too often these decisions are made more by circumstances and accident than by farsighted planning. Steven Johnson recounts the story of the Collect Pond, a freshwater body in lower Manhattan that in the early nineteenth century was nearly protected and turned into a park that could have been one of the world's great urban spaces. But conservationists were unable to mobilize sufficient support, and by 1812, the pond was filled in and developed into housing, eventually becoming the notorious Five Points slum.[53]

In climate policy, uncertainty is a major barrier to action. Consider, for example, efforts to decarbonize electricity generation, now considered a relatively low-hanging fruit of the mitigation process. In 1992, when the UNFCCC was created, non-hydro renewable energy accounted for a negligible fraction of global power generation. Just a few places in the world had wind turbines or solar panels, which were vastly more expensive than fossil fuel plants. Even where governments were willing to begin investing in these options, grid operators worried how these kinds of intermittent sources could be integrated at scale and so saw a need for stable "baseload" generators (like coal or gas plants) that would smooth the gaps between sunny and windy days. Even as the cost of solar and wind plummeted to reach parity (or beyond) with fossil fuels, this worry about how to incorporate new technologies into old grids has continued to be a major barrier to speedier rollout of renewable energy.[54]

Finally, uncertainty around long problems also follows from our agency to shape them. Long problems are uncertain in part because they are indeterminate. This openness means that decisions made early in a problem's temporal frame can lead to starkly different outcomes. After all, nothing is set in stone. The previous section noted that political institutions often exhibit path dependence, but so too do other aspects of problem structure. For example, at the time of writing, global emissions will need to fall something like 7 percent each year until 2030 to reach the 1.5°C pathway outlined in the IPCC's 2018 report. Imagine, however, if we had managed to peak global emissions in 2012, on the twentieth anniversary of the UNFCCC, and had begun reducing from then. The annual reduction needed to achieve the same result in 2030 would only be 1–2 percent per year because our peak would have been lower and we would have more time to bring it down (see chapter 6). It still would have been an enormous task but a much more tractable one. Now imagine we delay action by another decade. How much harder will it be to achieve climate safety then?

Although uncertainty is a key barrier to early action, it is not the only one. Sometimes problems are well understood and action still falls short. Consider infectious diseases with pandemic potential. Experts had predicted the emergence of a threat like COVID-19 for decades. Commissions, ranking systems, and other measures were set in place to understand and prepare for its inevitable emergence, which were bolstered by more limited outbreaks like SARS or Ebola. Although of course we could not predict exactly when a pandemic would occur and what characteristics the disease would have, we knew such threats were coming. Despite this relative certainty, our preparations were manifestly insufficient.

Alongside uncertainty, distant effects are difficult to mobilize significant political action around because they lack salience. Even when information is available, there is no guarantee that actors spend their scarce attention on it, as discussed at the start of this chapter. Policymakers can address only so many topics at once, forcing the most urgent to the top of the priority list. Interest groups and citizens, also needing to prioritize urgent matters, lobby governments to address their most pressing issues. When effects sit far in the future, it is difficult for governments or those that influence them to take attention and resources from urgent priorities to the important things that no one is talking about.

The combination of uncertainty and low salience, in turn, enables obstructionism, the ability of interests tied to the status quo to maintain their interests. Consider the hurdles a policy entrepreneur would have to overcome to create and implement a policy addressing a problem with distant effects like climate change. First, that policy entrepreneur would have to herself see value in pursuing an obscure issue, one that is unlikely to garner her a quick win and the associated political benefits. Few will have incentives to pursue such causes. Second, she would have to mobilize a sufficient coalition of interests to be able to influence policy. This would require each of those interests choosing to focus on a distant topic over their more urgent priorities. Third, this interest coalition would need to force the issue onto the broader political agenda, competing for limited space with numerous immediate priorities. Fourth, the coalition would need to somehow overcome, compensate, or neutralize political opponents.

To the extent those opponents are worried about the short-term costs of action, everything that is hard for the long-sighted policy entrepreneur will be easy for them. Opposing long-sighted policy—that is, promoting short-term outcomes—will give them the opportunity for quick wins on issues that are relatively easy to mobilize interests around. And even if the long-term-oriented policy entrepreneur wins a battle, she must preserve and maintain those gains permanently, as opponents will seek to reverse any defeats they face. A one-off victory may be important, but long problems often require sustained policies over time, while it takes only one victory by opponents to block them. The longer a problem's effects reach into the future, the more friction the policy entrepreneur will face at every stage, and, should she get a win, the more enduring her victories will need to be.[55] For these reasons, obstructionism becomes a major political barrier to early action.

Technology development provides a good example of how uncertainty, low salience, and obstructionism block effective policymaking. New technologies can have radical and unintended consequences, but these are very hard to know when they are just emerging, and they lack salience. Once the effects of a technology are understood and widely felt, however, it may be very hard to change that technology as it now has producers and users who will resist efforts to rein it in. As David Collingridge described the "dilemma of control" in 1980, "when change is easy, the need for it cannot be foreseen; when the need for change is apparent, change has become expensive, difficult and time consuming."[56]

Stimulating technological transitions faces a similar but inverse problem, as uncertainty and low salience creates opportunities for opponents to block or delay action, facilitating obstructionism. If there are legitimate doubts about future outcomes, more study and experimentation is always justified, and the risks of going down the wrong path cannot be ignored. Under such conditions, experimentation is a sound policy (see chapter 3), but as policy experiments unfold, the inevitable failures (a 100 percent success rate is a sure sign of an ineffective innovation policy) can bring political costs. Famously, the US solar company Solyndra received a $500 million loan from the Department of Energy in the wake of the 2009 financial crisis to support the development of an innovative photovoltaic technology. But shortly thereafter, the cost of the alternative technology, silicon-based photovoltaic cells, fell dramatically, making Solyndra's product unviable. The company went bankrupt in 2011. It was also revealed that Solyndra had misled the government in its loan application. Fossil fuel interests seized on the case, and Republicans in Congress made it a centerpiece of their efforts to block further support for renewables, even though, on average, solar technology has been extraordinarily successful.[57]

Under these conditions, muddling through becomes the more likely outcome. Unfortunately, a series of small incremental choices will struggle to deeply affect a long problem. Only when these incremental steps are additive and show increasing returns can they hope to develop into an effective response over time.[58] The history of renewable energy is in fact such a case, demonstrating how small investments and experiments by leaders can, over time, create technological changes that shift the problem structure in a positive direction. I return to such "catalytic" strategies in chapter 3. But other aspects of climate and other long problems do not share this positive dynamic, unfortunately. For example, current investments in adaptation are woefully below optimal levels. Even in wealthy places where resources are sufficient and damage estimates are relatively clear, we are not building the seawalls, changing the zoning laws, restoring the marshes and mangroves, developing the irrigation systems, or protecting the groundwater resources we know we will increasingly need in the decades to come.[59]

Or consider negative emission technologies. Many climate scenarios assume significant deployment of these tools, which do not yet exist beyond the pilot stage, in the decades to come (including all of the scenarios that informed the influential 2018 IPCC report on the goal of limiting climate change to 1.5°C). The scale of such technologies modelers envision we may need is vast. One study suggests a land area the size of Brazil may need to be given over to bioenergy with carbon capture and storage (BECCS), a process of burning biomass for energy and permanently sequestering the carbon released, effectively removing from the carbon cycle on a geologic timescale.[60] How this would be done technologically, economically, and politically remains hazy.[61] Moreover, the issue is highly charged—what about the other uses of that land, or the people who live on it—and essentially taboo in the UNFCCC process. Under such conditions, the early action needed to deliver the outcomes the models envision is hamstrung.

For a long problem like adapting to climate impacts or deploying negative emissions technologies at scale in the next decades, muddling through is risky because early action is what matters. Building infrastructure, restoring natural coastal defenses, replenishing aquifers, and even relocating communities are drawn-out processes. Developing and deploying negative emissions technologies to remove billions of tons of carbon from the atmosphere will take similar amounts of time. If we wait until the need for them is urgent, they will be too late. Only more costly, less effective solutions will be available to us then.

The politics of early action are hard. Even for very short problems, like the spread of COVID-19, governments often struggle to act sufficiently in advance, only lumbering into action after a crisis emerges and the best solutions are already off the table.[62] Long problems instead force us to address the question, how do we get a crisis-level response well in advance of the crisis? How do we achieve wartime mobilization before the first shot is fired? The next chapters examine solutions to the problems raised above, starting with the central dilemma of early action.

3

Forward Action

ADDRESSING THE EARLY ACTION PARADOX

In 1992, countries promised, in international law, to "prevent dangerous anthropogenic interference with the climate system." In the decades that followed the adoption of the UNFCCC, countries met again, and again, and again, but "dangerous anthropogenic interreference with the climate system" got worse. It has been suggested that doing the same thing again and again and expecting a different result is a definition of insanity.

Are we crazy to muddle through? Lindblom's famous 1959 article of that title, discussed above, outlines the "many imperfections" of muddling but argues that given the constraints policymakers face, we should understand it as the de facto system for addressing complex challenges and "not a failure of method for which administrators ought to apologize."[1]

Had Lindblom considered long problems, he may have found more cause for contrition in policymakers. Although "successive limited comparison" is indeed common, it is manifestly unable to address long problems where causes and effects span long periods. To address a long problem, we need to act early to have a hope of influencing the outcome we want to affect, be it carbon emissions, resilience, urban spaces, pension stability, or other goals. But mobilizing early action is politically hard because of uncertainty and low salience and the obstructionism they enable. In other words, we cannot muddle through the early action paradox. How can policy work around this dilemma?

For many, the answer is some form of "anticipatory governance."[2] The term is used in different ways but typically involves a purposeful effort to consider multiple possible future trajectories and to seek to devise interventions that steer toward better outcomes and away from worse outcomes. For example, the term has been used in science and technology studies to describe a robust process for deciding how to manage emerging technologies like nanotech or AI.[3] Leon Fuerth, a US public servant and diplomat turned academic, has proposed a complete set of "anticipatory governance practical upgrades" for the US federal government, arguing, "if we are to remain a well-functioning Republic and a prosperous nation, the U.S. Government cannot rely indefinitely on crisis management, no matter how adroit. We must get ahead of events or we risk being overtaken by them. That will only be possible by upgrading our legacy systems of management to meet today's unique brand of accelerating and complex challenges."[4]

Anticipatory governance provides a fitting label for institutions and strategies through which the early action paradox might be overcome. The key question, of course, is what can make it politically possible? I review three families of governance strategies below, seeking to elucidate the political conditions under which they do or do not work. I begin with ways to make information about the future known and salient, a key enabler for other anticipatory governance strategies. Experimentalism, the next set of tools, describes governance strategies that use iterative processes of trial and error to remove barriers caused by uncertainty. Finally, catalytic strategies aim to progressively shift preferences over time, providing a way to erode the obstructionism that so often impedes early action. Although the early action paradox means that decisive preventative action will be rare, under certain conditions, a more anticipatory approach can ensure that incrementalism need not be mere muddling through.[5]

Information and Foresight: Making the Future Known and Salient

A first step toward long-term governance is to understand and appreciate the future implications of near-term decisions. Indeed, knowing the future effects of current causes is needed before a long problem can even be identified. However, the impact of information about the future depends on how that information changes political incentives and behavior in the present.

Knowledge matters because for many problems, even not particularly long ones, uncertainty is a major barrier to effective policy. As the journalist and policymaker Bina Venkataraman points out, the residents of Roman-era Pompeii did not understand the sources of volcanic eruptions, so they were not able to make effective long-term urban planning decisions.[6] Had they known the long-term consequences of dwelling beneath Vesuvius, they would likely have moved elsewhere. Indeed, building codes today are different in earthquake zones than in other areas because we understand and can measure earthquake risk much better, albeit imperfectly.

In the realm of climate change, a vast machine of knowledge systems have combined to understand the problem, its effects, and potential solutions.[7] Policymakers rely heavily on integrated assessment models, complex computational models that estimate how the global climate will change as a function of different choices we make around human activities (see chapter 6). Thanks to decades of science, we have a clear picture of what the problem is and what needs to be done to solve it.

However, although changing the information available to decision-makers can make them aware of the existence of a long problem, allowing them to make better choices in line with their preferences, information alone does not change their underlying interests. As such, informational tools are helpful for empowering and aligning political actors who already seek to act in a way consistent with long-term goals (or at least those who do not oppose doing so).

To have weight, though, information cannot just be available; it must be salient. Given the limited attention of policymakers and the glut of information available, the mere existence of information cannot be assumed to have a political effect. Only when information about the future is salient does it have at least a fighting chance of informing the decisions of political actors. For example, a pro-environmental legislator may not have a strong view on supporting a particular infrastructure investment project. But if the project is subject to an environmental review that publicly lays out the positive or negative impacts of the project decades hence, the legislator or the stakeholders she is accountable to may gain an additional reason to support or oppose the project.

Importantly, salient information does not just help actors better understand their own interests. It can in some circumstances alter the balance of political power between actors. For example, when a certain long-term topic rises in salience, advocates gain informal agenda-setting power, giving them a greater ability to squeeze a topic into the limited decision-making space available. Salience can also make it more difficult for narrow interest groups to influence decision-making because it brings a wider range of interests into play.[8] A legislator may be happy to support a single company's position in the quiet politics of a backroom deal but may take a different position if forced to justify her stance to constituents and journalists at every public encounter. In these ways, changes in the information available about the future, and especially in the salience of that information, can affect the balance of power around decision-making in the present, largely by increasing the effort and ability of already aligned political actors in the present.

Salience can result from events. A major earthquake, financial collapse, pandemic, war, or other catastrophe can force policymakers to consider information about long-term risks and consequences. Sometimes shocks are so large that substantial changes result. For example, the 2008–2009 financial crisis drove significant changes in many countries' regulations, though many observers concluded they were in aggregate less than was needed to prevent further such crises.[9] Similarly, countries that had experienced large infectious disease outbreaks likes SARS in the decades before the emergence of COVID-19 performed better in the first months of the global pandemic.[10]

In the climate realm, researchers have found some evidence that certain kinds of climate impacts, particularly unusually high temperatures, can raise the salience of climate change in surveys and other measures of public opinion, though the effect has tended to be modest and is conditioned by prior beliefs and elite communication around events.[11] And public opinion is of course just one factor affecting policy outcomes. There is no systematic evidence that jurisdictions that have experienced greater climate impacts to date have adopted more vigorous mitigation or adaptation measures. These results may of course change as climate impacts grow more severe, but this lag, should it materialize, highlights the limits of this reactive mechanism.[12]

From a theoretical perspective, it is straightforward to see why "autocorrection"—when a crisis raises salience and engenders a response—may be helpful for recurrent problems but is inadequate for long problems. Such a mechanism can only work at least one crisis too late, like the immunity gained from infection. But for long problems in which consequences manifest significantly after causes, post hoc action comes too late. Compounding this difficulty, there is no guarantee that reactions to crises will create incentives for long-term strategies, such as preventing greater climate change through further reductions in emissions. Indeed, in the face of an immediate crisis, short-term protection may become the overriding imperative, such as responding to current climate impacts by building higher seawalls instead of reducing emissions. To the extent this is true, crises may drive more short-termism, not less, a plausible scenario sketched in chapter 6 and discussed in the conclusion. Moreover, the salience generated from crises may not endure after the crises recede. For example, a number of surveys show that the effect of extreme weather on public attitudes toward climate change recedes fairly quickly once the heat wave or storm passes into memory.[13]

For the above reasons, we cannot count on information or salience to emerge organically for long problems. Put another way, political attention will be systematically undersupplied for such challenges. We must instead create it. There are essentially two ways to make information about long problems salient: by the political strategies of actors in the present, particularly representation or activism (discussed in the next chapter), or by weaving future-oriented informational tools into decision-making. A first step toward overcoming the early action paradox is therefore to create more institutions that both help us understand the future and raise the profile of that information in politics and in the policy process. This can be done in several ways.

First, governments often have procedural requirements that force decision-makers to consider the long-term implications of policies, especially in budgeting and planning. Many writings on long-termism cite the Iroquois people's rule that decisions in the present should consider the impact across seven generations. In contemporary policymaking, there are innumerable examples of similar rules requiring decision-makers to look forward, though few have such an extended horizon.

One of the most common requirements to consider the future is the mandatory assessment of how spending measures and budgets affect the fiscal health of a jurisdiction. Such requirements are ubiquitous in some form because nearly every government that can borrows from capital markets to finance expenditure, commonly by issuing bonds to be repaid from future tax revenues over five, ten, twenty, thirty, or more years. In 2020, during the COVID-19 pandemic, government debt globally added up to 256 percent of global gross domestic product (GDP), a new high.[14] When capital markets collectively decide that a government will struggle to repay, the cost of borrowing goes up, limiting what governments can do in the present and potentially provoking a sovereign debt crisis. Present political leaders in nearly every government therefore have a strong, structural incentive to be able to credibly demonstrate longer-term fiscal health.

A prominent example of this kind of information provision is done by the Congressional Budget Office (CBO) of the United States, a specialized body created by Congress to advise it on the fiscal impacts of legislation. The CBO gives independent, nonpartisan cost estimates for proposed laws, showing how revenue and spending would change vis-à-vis a baseline should a new law be enacted. Such scoring is automatic—under law, nearly every piece of proposed legislation that comes out of a congressional committee triggers a CBO review—and highly salient. This salience derives from both the institutional authority of the CBO as an independent, technocratic agency of the US Congress but also because the fiscal position of the federal government is a major topic in American politics, as it is in nearly every country in the world due to government reliance on capital markets and the way taxation directly affects individuals in a significant way. If political actors, particularly on the political right, can cast their opponents' policies as fiscally irresponsible, they are much more likely to block new policies, giving the CBO significant power. Indeed, its judgments carry such political weight in Washington that the Harvard political scientist Theda Skocpol has described the CBO as a "nearly sovereign" actor in American politics.[15]

Contrast the CBO and its analogues around the world with another common example of requiring the generation of knowledge about the future in decision-making: the use of environmental impact assessments (EIAs) and related tools. EIAs seek to understand the consequences for the environment of certain projects or actions and to explore options for mitigating harms. The world's first EIA process was adopted by the US Congress in the 1970 National Environmental Policy Act, which included the goal to "fulfill the responsibilities of each generation as trustee of the environment for succeeding generations."[16] Since the 1970s, EIAs have become nearly ubiquitous in some form or another across the world, albeit taking many different forms across local governments, national policies, and the activities of international organizations.[17] Typically, the requirements of an EIA increase with the scale or scope of a decision, but in some places, even relatively small projects trigger environmental review. At one end of the spectrum, EIAs may be cursory processes carried out by the entity running a project or its hired consultants. Its results may be used only internally, taking the form of suggestions. At the other extreme, EIAs can be rigorous, complex, and expensive affairs carried out by regulators or independent agencies who make binding decisions on how projects develop. They can involve significant public participation and consultation, and their outcomes can become matters for scrutiny and debate, or the basis of legal challenges.

Governments have sometimes created institutions that perform similar functions on a larger scale. For example, the New Zealand Parliamentary Commissioner for the Environment was created in the 1986 with a broad mandate to review and report on the environmental impacts of the country's laws and regulations, either prospectively or retrospectively, and to propose ways to improve them.[18] More generally, many countries have included environmental or climate criteria in so-called regulatory impact assessments (RIAs), which assess forthcoming laws and regulations for a variety of impacts on sustainability.[19] In the United States, for example, cost-benefit analysis of regulatory decisions by executive agencies have been required to consider the "social cost of carbon," an attempt to quantify the future harm of climate change via integrated assessment models and then account for it in current regulatory decisions.[20] By making consideration of future impacts automatic, the social cost of carbon exerts significant influence on administrative rule-making. Although not particularly salient, this technical rule is important in the US regulatory process because administrative agencies' rules are often contested in court and need to be able to demonstrate that their cost-benefit analysis was robust to stand muster. One indication that this rule was seen as consequential was the fact that the anti-climate Trump administration removed it as one of its early acts in office, though it was reinstated by the subsequent administration.

Foresight is another and qualitatively different tool for generating information about the future. Foresight describes a family of approaches and techniques that provide heuristics for long-term planning and strategy.[21] It includes practices like horizon scanning, trend analysis, scenario analysis, and their various permutations and variants.[22] Common across the many flavors are first, a structured way of identifying future possibilities and contingencies and weighing their implications. These possibilities may arise from technical forecasts (e.g., from climate models or budget projections), extrapolation of existing trends, or simply imagination. Second, foresight practices seek to draw in diverse sources of knowledge to avoid groupthink and maximize the likelihood of identifying black swans and other unexpected outcomes. As political scientist Philip Tetlock's well-known studies of expert judgment show, experts often have a hubris problem; more accurate forecasters showed "a blend of curiosity, open-mindedness, and unusual tolerance for dissonance."[23] Foresight techniques attempt to overcome some of these limitations and cognitive biases in technocratic forecasting.[24] For example, the Delphi method, developed by RAND in the 1950s, involves a diverse group of experts making predictions regarding a topic and then anonymously sharing their predictions iteratively, challenging and learning from each other's conclusions.

In this way, foresight differs from forecasting, which provides a narrower informational assessment of future outcomes or impacts. For example, the scores assigned by the CBO or the scenarios studied by the IPCC tend to be more bounded and specific than a foresight process and are grounded in an established technical methodology. Although this kind of knowledge can be a valuable input for a foresight process, the latter tends to be more expansive and diverse, aiming to assemble knowledge beyond a single disciplinary lens or precise methodology. For example, it might consider a climate outcome generated from an integrated assessment model and attempt to develop multiple logically consistent narrative pathways for how that outcome might be reached or avoided. For many academics grounded in specific disciplines that create rigorous processes for how knowledge can be validated (or rejected), the looser nature of foresight makes it easy to dismiss as mere prognostication (see chapter 6). But given the widespread and growing use of the practice, a more interesting question is what effects it may have on policy processes.

From its origins in the RAND work of the 1950s, foresight has become a relatively common practice across the public and private sectors. Many organizations now incorporate various foresight-related exercises into their planning processes, often drawing on a cottage industry of specialized consultants and facilitators. These foresight professionals can pursue degrees and certificates in the field from dozens of universities around the globe. Moreover, although a growing number of practices are explicitly labeled "foresight," the basic practices— identifying future trends and possibilities and attempting to devise ways to steer in a desired direction—are relatively ubiquitous, typically implicit in nearly all strategic planning or complex problem-solving exercises. In this sense, "foresight" can be understood as a range of ways to augment, regularize, and make more explicit long-standing organizational approaches to planning.

In policy work, many governments at different levels have a range of strategic planning functions that draw to some degree on foresight.[25] Perhaps the most common area for their use is in security and intelligence, where strategic planning and thinking through alternative possible futures is regularized across the world.[26] But foresight or similar activities can be found across many areas of government around the world, including economic development, public health, environmental governance, technology governance, crime prevention, and many others.[27] These may exist in a single ministry or across several, at national or subnational level, in the executive or the legislative branches. Such processes may be institutionalized or ad hoc.

Finland is commonly held up as the policy system in which foresight has become most mainstream.[28] Since 1993, the Finnish parliament has had a Committee for the Future, a cross-party group of sitting legislators tasked with grappling with the longer term challenges facing the country. This includes commissioning reports on longer term topics (e.g., technology, climate change, demographic shifts) and responding to the government's regular Report on the Future, which the prime minister's office prepares through a cross-ministry working group each electoral term. All ministries support this work through the joint Government Foresight Group, which also works with the National Foresight Network, a broad collection of stakeholders from the private sector, civil society, academia, and other sectors. This National Foresight Network encompasses a wide range of local governments, civil associations, and private sector entities that both incorporate foresight into their own work to varying degrees and also collaborate nationally. In this way, the entire Finnish policymaking apparatus is engaged on a regular basis in thinking through potential future pathways for the country. Although elements of this system exist across many other countries—and it has been proposed in many more[29]—in Finland, it is perhaps most advanced.[30]

One of the benefits of institutionalizing informational tools and foresight practices as ongoing processes—as opposed one-off reports or commissions—is that they almost always get the future slightly wrong and sometimes very wrong. At the end of the nineteenth century, major cities that depended on horses to get around were slick with their manure. As authorities struggled to find a solution, the prognosis looked grim. One New York observer feared that by 1930, horse dung would block out the first two stories of the new Manhattan skyscrapers. In fact, New York's last horse-drawn streetcar closed in 1917 as the automobile took over.[31]

Systems that provide information about the future therefore need ways to manage and deal with their inevitable errors. Forecasts like those from climate models tend to come with uncertainty statements and caveats, but many users lack the knowledge to gauge them appropriately. Other tools like foresight processes, though not meant to be predictive, may omit key trends or perspectives that impair their ability to generate helpful ideas on possible future pathways. Wrongly conceived, informational tools may even lead to false certainties that can even become self-fulling prophecies. For example, if an intelligence assessment of an enemy's intentions conclude that they will almost likely attack in the future, decision-makers may decide to preemptively strike, creating a conflict that could have been avoided.[32] Making information about the future known and salient is necessarily imperfect, and therefore tools that seek to do so require continual challenge, updating, and revision.

Once knowledge about the future is created through these kinds of informational tools, does it actually change things? Determining the precise impact of EIAs, RIAs, foresight mechanisms, and similar instruments is challenging, given the difficulty of establishing a relevant counterfactual for each case. The causal chain from epistemic benefits to action is rarely directly observable.

Even if the mechanisms are relatively easy to specify (changing policymakers' information or perhaps even preferences), it will be only one of many causal chains leading to a given outcome. For example, Finland perhaps represents a most likely case for impact, but empirical assessments of its foresight system have struggled to identify decisive policy outcomes that would not have happened but for the anticipatory framework.[33]

Cognizant of these inferential challenges, the rough consensus is that informational tools for environmental protection in particular have an incremental impact on top of other environmental regulations. The most common outcome of EIAs is a "relatively modest fine-tuning of developments," according to one synthetic analysis of EIAs across many jurisdictions.[34] Similarly, in contrast to the "nearly sovereign" US CBO, the New Zealand Parliamentary Commissioner for the Environment has had a "positive" track record in having its recommendations adopted, according to Boston, with 60–70 percent of its recommendations being adopted. However, he adds, "while such results appear to be impressive, it must be recognized that recommendations vary greatly in their significance and that Commissioners are unlikely to make repeated recommendations of a kind that they know will be openly ignored."[35]

Because of the early action paradox, EIAs and RIAs, budget scores, foresight processes, or other kinds of institutions that generate information about future impacts matter when they help mobilize political actors already disposed to act in the long-term interest or bolster the ability of those actors to set the agenda or otherwise enhance their influence. Put another way, they strengthen the ability of political interests in the present to advocate for long-term outcomes, provided their interests are already compatible with those long-term interests. Understanding this critical scope condition for these kinds of tools helps identify the conditions that make such procedures more or less effective and helps explain the relatively greater effectiveness of fiscal checks over environmental checks.

For informational instruments to matter, actors that hold or are influenced by long-term interests must both exist and wield some modicum of power. In jurisdictions with an active civil society, the rule of law, a vibrant media, and political contestation, environmental groups can often use information disclosure as a formidable political asset. The New Zealand Parliamentary Commissioner for the Environment, for example, can benefit from these conditions, but an analogous institution would not be expected to be as effective in contexts where those conditions do not hold.[36]

By the same token, when information about the future is generated automatically and publicly, it has a stronger chance of being found and picked up by some political actors willing and able to use it. When EIAs are public, they stand a chance to mobilize at least some groups concerned with a given environmental impact. But if the results are private or unknowable, then there is no possibility for opposition to coalesce. Moreover, when institutions are highly competent and respected, the information they produce gains additional authority. For example, in the New Zealand context, the Parliamentary Commissioner for the Environment has cultivated a reputation for technical precision and effectiveness and so is accepted as an authoritative voice on environmental matters.[37] When it weighs in on a topic, its very standing makes the information salient.

Similarly, institutions that provide information regarding future fiscal impacts like the CBO will matter little unless there are actors in the present who care about that information and mobilize around it politically. But in contrast to environmental information, fiscal information has a structural advantage in that capital markets automatically assign some degree of value to it, changing the cost of borrowing, and therefore immediately influencing political incentives in the present. This means that any opponent of a new law or regulation that the capital markets deem to negatively impact the future fiscal balance of a country is immediately empowered politically.

The contrast between the political heft of future information about the environment and that pertaining to the fiscal health of a country shows the limits of information as a way to achieve long-term governance. When information about the future can mobilize actors in the present, it can have some political impact. But the extent to which it can do so is highly conditioned by the balance of power and interests in the present. When allies of future interests are strong, the information can matter significantly. When they are weak, it is unlikely to significantly alter outcomes.

Compounding this difficulty, political actors in the present represent future interests in a way that advances their short-term political goals. Unsurprisingly, actors in the present use arguments about the future—like any kind of arguments—opportunistically and inconsistently. For example, it has been observed that conservatives are more accepting of budget deficits that result from cutting taxes than those that result from increasing spending, since they tend to oppose greater social spending irrespective of its fiscal impact. Writing from a critical theory perspective, Liam Stockdale notes how countries have invested enormous resources and effort and restricted civil liberties in the name of preventing future terrorist attacks but by and large do not show commensurate efforts to prevent climate change.[38] These examples show how future interests are likely to be picked up selectively and unevenly by actors in the present. As information about the future becomes known and salient, those future interests of greatest value to current actors will gain allies in the present. But those future interests that attract only scant attention from weaker actors will remain at a disadvantage.

For all these reasons, informational mechanisms go only so far. Financial crises occur with disturbingly regular frequency, despite expert warnings and prevention strategies.[39] Pandemics like COVID-19 have been anticipated for decades, and yet the world was still largely unprepared to manage it. Making information known and salient is only the first step for overcoming the dilemma of early action. It is therefore probably best to think about informational tools as key enablers of early action, necessary but typically insufficient. Instead, we should think about how to combine them with other strategies for addressing the early action paradox, which the following sections address.

Experimentalism: Overcoming Uncertainty

Uncertainty is one of the key drivers of the early action paradox. If we do not know exactly what technologies will replace our carbon-based energy system, it is difficult to make the investments needed to effect that shift. If we do not know exactly when and how climate change will affect us in the future, we do not know what to do to adapt to its impacts. Uncertainty of course is not limited to long problems like climate change, but as chapter 2 argues, it correlates with problem length for the simple reason that all else equal, more change is possible in longer stretches of time, and the impact of an action in the present has a longer, more complex causal chain to the outcome of interest. Long problems therefore require techniques to tackle uncertainty.

As the mantra goes, "When in doubt, figure it out." Experimentalist governance provides a way to iteratively test and scale up solutions. Through flexibility, trial and error, learning, and diffusion, it tackles uncertainty directly by making the *process* of solution-finding the solution. This approach helps avoid the political barriers to action that uncertainty generates because it can reduce uncertainty down to a level where decisive action is more possible. As political scientists Charles Sabel and David Victor argue in their 2022 book on the subject, experimentalist governance "explains how firms, public authorities and civil society institutionalize the coordination and learning that complexity makes necessary without attempting the kind of planning that uncertainty makes impossible."[40] Although experimentalism works best under certain conditions and is not a panacea for addressing the obstructionism that often blocks action on climate change and other long problems, it can be a useful tool for making progress on issues where sufficient political support exists to create shared goals and a robust process for pursuing them.

Experimentalism in governance has a long history and many guises. The early twentieth-century political scientist John Dewey's theory of "pragmatism" argued for delegating decision-making authority to the people and places with most direct experience of a problem and iterating and learning from the way they addressed it. He argued for a participatory system of joint problem-solving between experts and people living with problems, famously stating that "the man who wears the shoe knows best that it pinches and where it pinches, even if the expert shoemaker is the best judge of how the trouble is to be remedied."[41]

These ideas have influenced a variety of thinking on governance. For example, adaptive management has become an organizing framework for much of modern environmental governance. In this approach, goals and strategies are iteratively discussed and reviewed alongside changing conditions. Regulators, regulatees, and a wide range of affected stakeholders share information and perspectives on how to best address a challenge. A related idea is reflexive governance, which similarly advocates for ongoing evaluation and iteration around a problem through the synthesis of expert knowledge and the perspectives of involved and affected interest groups and communities. Scholars of sustainable development have argued that complex interaction between human and natural systems "has brought with it recognition of the limits of rigid analysis and the inadequacy of policy approaches that aim at planning and achieving predetermined outcomes."[42] Instead, they advocate for a governance system that incorporates several features designed to promote flexibility and adaptability.[43] Chapter 5 returns to this broader challenge of institutionalizing durability with adaptability.

Experimentalism as an organizational strategy is also linked to changes in the private sector during the late twentieth century.[44] As globalization and related trends increased market uncertainty, firms that were too slow to adapt faced losing out to more nimble competitors. Successful firms therefore moved from hierarchical planning to rapid adaptation. To achieve this, decisions were pushed from central leaders to a multitude of more flexible, operational teams tasked with figuring out the best way to adapt to new market conditions. Central decision-makers were responsible for defining (iteratively) objectives, reviewing performance, and convening the various units to learn from each other, but many substantive decisions on what to do and how to do it were increasingly decentralized.

Interestingly, organizational patterns observed in the world's largest capitalist firms can also be seen in the Chinese Communist Party. The expression "crossing the river by feeling for stones" has become a cliché of Chinese economic policymaking, but trial and adjustment are hallmarks of the Chinese system. As Sebastian Helimann puts it, "China's governance techniques are marked by a signature Maoist stamp that conceives of policy making as a process of ceaseless change, tension management, continual experimentation, and ad hoc adjustments."[45]

Among this family of experimentalist governance approaches, Sabel and Victor provide a particularly clear and distinctive model. In their conceptualization, experimentalist governance has five features. First, a goal is defined as well as metrics to evaluate progress toward it. This goal is typically set through a political process such as intergovernmental negotiations, legislation, or a regulatory decision—for example, the goal of reaching net-zero emissions by 2050. Second, the various entities responsible for implementing the goal pursue efforts to achieve it as they see fit. These entities could be countries, local governments, firms, or other types of organizations. Indeed, any actor with the capacity to develop and implement some kind of solution can be involved in an experimentalist system in some way. Third, as a condition of their autonomy, the various implementing units must report results and share experience through rigorous peer review processes. This step is critical to allow for learning but also to enable the fourth element: iterative revision of goals and metrics in light of experience.

Fifth and critically, Sabel and Victor argue that experimentalist approaches require not just flexibility but also, ultimately, coercion, a feature they term a penalty default. Unlike a conventional regulatory sanction, actors that are making good faith efforts to implement the goal are not initially penalized. Indeed, those that lack key capabilities to implement a goal may be given additional resources to help them adopt the new technologies or practices required. Only once uncertainty around what to do and how to do it has reduced, or if actors do not engage in the good faith efforts to develop solutions, do penalties apply. Conventionally, penalty defaults can take the realm of normal regulatory punishments like fines or bans. But penalty defaults may also result from informal or "soft" governance measures like reputational costs.

Sabel and Victor argue that experimentalist governance approaches are far more common than is conventionally understood, particularly in the environmental realm, holding up the 1987 Montreal Protocol to address ozone-depleting substances as one key example. Montreal is widely seen as one of the world's most successful environmental treaties but for the wrong reasons, Sabel and Victor argue. The conventional wisdom is that Montreal (and successor agreements in the ozone regime) succeeded by setting strict targets and timetables for phasing out harmful pollutants. Failure to meet these legally binding targets could trigger trade penalties, although developing countries benefitted from a longer phaseout and a funding pot to assist them to adjust. This model, which Sabel and Victor term the Montreal Mould, strongly influenced the institutional design of the 1992 UNFCCC and especially the 1997 Kyoto Protocol. Unfortunately, this interpretation of Montreal misses the experimentalist features that, in Sabel and Victor's argument, actually made it successful, which they term the Montreal Method. These elements operated through a series of dynamic technical bodies working on specific aspects of the problem. Whereas high-level targets were set "top down" through negotiations, the detailed rules of when and how specific substances were phased out were heavily influenced by more dynamic groups of industry and regulatory experts who would review collective progress and make adjustments as needed.[46]

As the example of Montreal shows, experimentalism can be a powerful tool for addressing uncertainty and therefore overcoming the early action paradox around long problems like climate change. When can this approach work, and when can it not? Several necessary or enabling conditions follow from the causal mechanisms it posits.

First and most obviously, a diverse array of actors must be willing and able to do the kind of rapid experimentation that can create solutions. This is often the case for problems amenable to the development of new technologies and business models, such as creating alternatives to ozone-depleting substances or deploying zero-emissions vehicles. For these sectoral transition challenges, there are typically a number of firms that create, sell, and buy the relevant technologies, as well as a range of governmental actors who shape the market as consumers or regulators. Each of these actors can be mobilized into an experimentalist framework. For example, in the climate realm, there are dozens of sectoral initiatives at the national and international levels in areas like electric vehicles, green hydrogen, zero-emissions steel, shipping, aviation, and many others.[47] Such initiatives tend to set a specific target and then promote some kind of collaboration among industry and/or government to innovate toward it (though most lack penalty defaults).

Given that decarbonization requires technological innovation of this kind, this first condition for experimentalism applies widely to the climate challenge. Note, however, other long problems, including other aspects of climate, do not have this same array of actors that can be mobilized to experiment. Imagine a city making a long-term infrastructure investment to adapt to climate change, such as building a seawall, or a government seeking to address a generational shift in the ratio of workers to the elderly. In both cases, there is only one actor and only one policy decision at any given moment in time. The city cannot try out a seawall, a managed retreat and relocation program, and a mangrove afforestation project and then see which works best to tackle sea-level rise. It gets only one chance (though of course cities facing later sea-level rise could learn to some degree from those on the front line). Similarly, there is not a multitude of actors looking to design the best state pension system; only the government can do that. Experimentation works best for problem structures in which many actors can affect the problem and solutions can be rapidly developed and iterated.

Second, for experimentalism to work, there has to be the possibility for learning and diffusion. This requires a process in which actors' solutions are robustly evaluated, with the most effective approaches becoming wildly adopted. Again, technology-focused innovation challenges are particularly ripe for this model. Learning can be advanced through public and private research and development spending. Different approaches can be evaluated by precise performance metrics. Diffusion can then occur through both markets, as the more successful models outcompete others and through government incentives that shape markets. If the experimentalist approach can be appropriately institutionalized by a capable and technically sophisticated regulator in a way that enables real review and evaluation, learning and diffusion follow.

For many complex challenges or long problems, these conditions will obtain, but for others they will not. For example, it may not be obvious which solution is better or worse because they involve not just technical performance standards but complex trade-offs and value judgments. Additionally, diffusion may be impeded by a high degree of context specificity. For example, land restoration and preservation projects are an urgent part of addressing the climate challenge as well as global biodiversity goals. But a solution that works for a rainforest in Colombia may not suit one the Congo, and a successful Congolese model may not work in Indonesia.[48] Land use policies are significantly shaped by local environmental, social, economic, and political conditions, which can vary substantially from place to place, limiting the scope of diffusion.

Third and perhaps most importantly, the experimentalist approach requires a certain degree of political consensus around, or political power in, the regulatory authority. Political consensus and/or power is needed for two of the key elements the experimentalist framework requires: goal setting and the imposition of a penalty default. Experimentalist governance does not simply emerge organically. Authorities need to decide to embark on a project of experimentation toward a common goal, even if that goal will be refined and elaborated through the process. If there is firm opposition to a certain outcome, experimentation may never get going.

This scope condition applies perhaps even more strongly to the imposition of a penalty default. In the experimentalist framework, the credible threat of penalties is necessary to ensure ultimate alignment around the objective but also to ensure robust participation in the experimentalist process in the first place. If interest groups know the status quo will not remain stable, they have a strong incentive to participate meaningfully in the process of finding new ways forward. Unless the authority has the power to impose real costs on those that do not join the experimental approach, the penalty default will remain weak.

For these reasons, an effective experimentalist governance system is predicated on a problem structure where the balance of power and interests favors *some* solution, even if uncertainty prevents actors from converging around *which* solution to go with and thus holds back action. In other words, it is well positioned to address the problem of uncertainty but is less effective against sheer unwillingness. However, even in the face of some opposition to goals or penalties, an experimentalist approach may be more palatable to adversely affected interest groups than a conventional "command and control" regulatory measure as the costs could be more uncertain and distant. In this way, experimentalism could emerge in part as a compromise or interim measure when political consensus or authority is only partial. Of course, for a long problem, it is critical to think through how the process of experimentalism may change the problem structure over time. The next section takes up this consideration.

These three necessary/enabling conditions—many actors able to iteratively develop solutions, capacity for learning and diffusion, and political consensus/power around goals and penalties—help explain why, through the experimentalist lens, the Paris Agreement is at best a modest instrument in Sabel and Victor's analysis. The first condition is relatively well satisfied, as the Paris Agreement encourages all countries but also cities, businesses, and a multitude of other actors to advance its goals and allows maximum flexibility to experiment with its nationally determined contribution (NDC) structure. This avoids the "targets and timetables" trap of the Kyoto Protocol. The second condition, however, is only weakly met, as the Paris Agreement's review mechanisms are very high level and abstract. Countries must report on progress toward their goals, but review and evaluation are fairly superficial as countries have resisted exposing themselves to meaningful scrutiny. There is a regular stocktaking of overall progress, but it occurs only every five years. The review mechanisms therefore serve more as high-level political processes and not as granular and specific tools for enabling learning and diffusion. Underneath this overarching architecture, the UNFCCC enables or encourages a number of more precise learning and diffusion processes such as the Technical Expert Meetings and the Marrakech Partnership for Global Climate Action, but these tools remain relatively small in scale to date. Most of the substantive work of review happens in private or multistakeholder initiatives, which may or may not dock into the UNFCCC. Finally, the third condition is also only weakly met. Although the UNFCCC has been able to set out long-term global temperature goals, there has not been sufficient consensus to agree on more specific sectoral targets around which more granular experimentation could be directed. Nor is the UNFCCC able to impose strong penalty defaults, though its goals give activists or the climate vulnerable a rhetorical hammer to punish laggards. So even though the Paris system has some features of the experimentalist framework, it remains too abstract to drive concrete learning and too sluggish and bureaucratic to drive rapid iteration; in addition, it lacks political support to impose penalty defaults. In the next section, I return to this assessment from a wider perspective and show how, despite the weakness of these features, other elements of the Paris system exemplify a logic for progressively overcoming obstructionism.

Overall, the experimentalist governance perhaps works best for issue domains where there is a degree of boundedness, such as the technological transitions in certain industrial sectors. Happily, these scope conditions include large swaths of the mitigation challenge, like transitioning heavy industry and transport to clean technologies. Even here, obstructionism may limit the use of experimental techniques. We therefore need to think more broadly about the conditions under which incentives in the present can be shifted over time. The next section picks up this challenge.

Catalysts: Eroding Obstructionism

Informational and foresight tools allow policymakers to look ahead and make choices around potential future outcomes clearer and more salient. Experimentalism creates a process for addressing uncertainty when there is sufficient political support and a multitude of experiments can iterate and successes can diffuse. Both tools help with the early action paradox of long problems, tackling information and uncertainty barriers. A final set of tools, catalysts, go further and aim explicitly to shift actors' preferences or the overall problem structure over time. Doing so has the potential to address the early action paradox most thoroughly because even though they grow incrementally, catalysts can initiate action in advance of consequences being felt, helping to erode obstructionism.

An influential article on addressing climate as a "super wicked problem" outlines the logic of such approaches. The object of policy, the authors argue, is to "trigger sticky interventions that, through progressive incremental trajectories, entrench support over time while expanding the populations they cover."[49] Instead of looking at path dependence only as a post hoc causal explanation, they argue we can seek to trigger processes that will generate path dependencies that steer toward a desired outcome. This self-reinforcement can occur through various additive mechanisms: locking in certain choices or institutions that are hard to reverse, actions that become increasingly costly or attractive over time, or actions that spread to larger and larger populations of actors. For example, a coalition might grow from a small group of early advocates to a wider set of interests, a new technology could create a new set of economic actors with different preferences to industry incumbents, or a norm could grow from a fringe belief to a common preference across society.

Such strategies often target "sensitive intervention points," which aim to realize the "potential for purposeful intervention to drive nonlinear amplification in complex systems."[50] Under what conditions can small scale interventions generate longer term, larger impacts? First, when a system is chaotic or near a critical point, a small kick may be sufficient to put it on a new path. For example, by subsidizing renewable energy to the point where it becomes economically attractive, policymakers can reshape the incentives that market actors face. Second, a shift rewires a system to behave in a fundamentally different way. For example, by requiring full disclosure of companies' and investors' climate-related risks, markets may alter the relative value of fossil fuel assets in ways that fundamentally reshape political power.

This logic for policy interventions can help overcome the early action paradox not only by altering information and salience or by reducing uncertainty (though they may do that to some extent) but also by chipping away at obstructionism. Recall from chapter 2 how mobilizing political action to prevent distant impacts faces a multitude of barriers, while those wishing to prevent action, obstructionists, find it easy to block and delay. Catalytic strategies that progressively change preferences and problem structure over time can erode the power of obstructionism by building up a critical mass of action that triggers path-dependent processes.

In the realm of climate mitigation, we can identify several dynamics through which policy interventions can take on a self-reinforcing quality that can make them catalytic.[51] As global environmental politics scholars Steven Bernstein and Matthew Hoffmann note, decarbonization in one area can scale and entrench in other areas through normalization, capacity-building, and coalition-building.[52] Material, informational, normative, and political economy mechanisms can be identified, with many of the first two overlapping with parts of the experimentalist approaches described in the previous section.

First, some climate policy interventions can affect the material costs around future interventions by changing technology and the economic systems around it. The literature on sociotechnical transitions shows how niche innovations by pioneers can diffuse to eventually reconfigure larger systems.[53] As technologies are developed and deployed, their costs descend down a "learning curve," becoming cheaper as more research and development is conducted and as production and distribution systems "learn by doing" and reach economies of scale.[54] Experimentalist techniques like those discussed above may be one way to trigger these kind of dynamics, but other tools can also work—for example, subsidies. Renewable energy technologies fit this pattern well. The cost of photovoltaic cells dropped 75 percent from 2010 to 2015 and wind turbines 30–45 percent in the same period; both technologies are now at or below the cost of fossil fuel alternatives in many parts of the world, sharply altering the incentives for taking climate mitigation action. These cost reductions were only possible because of costly support by first movers such as California, Denmark, and Germany.[55]

In addition to learning curves, many new technologies and business models demonstrate network effects; like telephones or email, the more people have them, the more useful they are. Consider electric cars or solar rooftops. For the first movers, such products are very inconvenient because there are too few charging stations, maintenance technicians, or electric cables and pricing systems that allow homeowners to send power back to the grid. As market penetration increases, however, the enabling environment shifts and later adopters are well supported. Indeed, once network effects are strong enough, they may help to lock in new technologies as default options.

Second, catalytic action can also generate learning effects. Again, the experimentalist strategies discussed above demonstrate this clearly, but learning effects can apply to "soft technologies" like governance methods as well. Just as mitigation actions generate new technologies and business models that alter material costs, so too do they produce new knowledge about policy design and implementation through experimentation and diffusion, further reducing costs and enhancing benefits.[56] This effect is significant because many areas of mitigation involve complex policy instruments. Emissions trading systems are a prominent example, where even highly competent regulators like the European Commission have struggled to make their credit allocation and exchange systems operate smoothly. For this reason, the Chinese government has drawn on significant international expertise in the development of its national emissions trading system and proceeded incrementally by first experimenting at the provincial level. In this way, previous actions generate epistemic resources that improve the ability of followers to emulate leaders.[57]

Third, over time, catalytic interventions may change norms around climate action, altering not just costs and benefits but also how actors value and prioritize climate policy outcomes. International relations scholars have shown how "norm entrepreneurs" engage in "strategic social construction" by attempting to shift norms toward their policy goals.[58] In the realm of climate, Bernstein and Hoffmann argue that actions in one sphere, such as a city setting a climate target, can "normalize" low-carbon preferences in ways that spill over to other actors.[59] These models follow a logic of increasing returns. Norms progress through a life cycle from emergence to a "norm cascade" in which they become widely followed in practice, to internalization in which they are embedded in the beliefs and preferences of most actors.[60] As more action takes place, the more this self-reinforcing logic applies. Although decarbonization norms are not currently widespread, some areas of climate mitigation show evidence of norm cascade dynamics, such as divestment from fossil fuel companies.[61]

Finally, catalytic interventions can rewire political economy, affecting the political processes of preference formation for states and other actors by generating new constituencies for action.[62] As new technologies emerge and grow, their producers and consumers develop a distributional interest in their continuance and expansion. At first, these new interest groups are unlikely to be able to overcome established incumbents in political contestation. But in economic sectors or geographic regions where incumbents are relatively weak, the new entrants may thrive and as action spreads, eventually acquire the size and clout to become politically competitive with incumbents in more and more jurisdictions and industries.[63]

Regardless of the mechanism—material, normative, informational, or political economy—catalytic interventions require certain conditions to be feasible. Principally, some form of path dependence must apply. This can take several forms: lock-in, increasing benefits associated with a certain course of action, increasing costs associated with not following a certain course of action, or an expanding population that adheres to the intervention. Moreover, these kinds of interventions will be more promising under certain structural conditions—for example, when a system is near a critical point.

Not all long problems exhibit these features. The previous section noted that experimentalism may not work as well for climate adaptation or protecting nature as it does, for example, the energy transition. Similarly, catalytic strategies may not find as many self-reinforcing dynamics in these areas. Building a seawall or moving a vulnerable population in one place or at one time does not make it significantly easier to do so again elsewhere, though perhaps it normalizes the practice to some degree. Planting new trees in one part of the world does little to catalyze action elsewhere. These scope conditions mean that for many long problems, catalytic interventions may be of limited value.

Beyond individual policy interventions, catalysts can also be institutionalized in various ways. We can define catalytic institutions as those that seek to achieve their policy goals by shifting actors' preferences or the problem structures they face over time. Many such institutions exist across multiple policy domains, but the Paris Agreement stands out as a particularly clear example of the catalytic logic (although, as I explain below, not all of its catalytic features are particularly strong). Catalytic institutions typically share one or more of the following elements, all of which can be found in the Paris Agreement.

Consider, first, goal-setting. Goal-setting can be a way to ensure durability (see chapter 5), but it also can be a way to influence preferences over time. Goals lay out high-level objectives and, in this sense, are statements about what a community or organization values and prioritizes. This can give them a normative quality, particularly when tied to a process that grants them legitimacy. For example, the SDGs, or the goals of the Paris Agreement, set out objectives agreed by all nations. Strikingly, these goals have then been taken up by a wide range of other actors—cities, companies, and so forth—as the appropriate targets for them to aim at. As many observers have noted, absent concrete plans and accounting, goal-setting can be cheap talk or even intentionally dissembling.[64] But when actors' preferences are shaped by a logic of appropriateness, goal-setting helps to clarify and crystallize what appropriate behavior entails (see chapter 5). For long problems, the ability of goal-setting to influence preferences through a logic of appropriateness can gain additional power to the extent it contributes to norm cascades, as described above.

Second, catalytic institutions seek to mobilize first movers through soft encouragement and flexibility and mobilization. Because the catalytic logic depends on initiating path-dependent processes, building up a small group of first movers is critical, even if they are marginal actors or if the steps they take are merely incremental. What matters is getting action going to initiate self-reinforcing processes. Catalytic institutions therefore seek to stimulate first movers to come forward, and to encourage small steps even from recalcitrant actors.

Flexibility is key to this logic. If a rule is too rigid—for example, requiring costly actions from countries or companies in the near term—obstructionists may block it entirely. By lowering the cost of action to what actors are willing to do, flexible frameworks help ensure that at least some action emerges. Moreover, flexible commitments allow the most pro-mitigation actors to put forward ambitious commitments, instead of limiting themselves to a least common denominator outcome. The Paris Agreement of course embodies this logic by shifting from negotiated commitments to NDCs.

Flexibility can also be about expanding the pool of actors. A further governance strategy commonly used to catalyze first movers is orchestration, the process through which actors like states or international organizations mobilize "governance intermediaries"—for example, cities, businesses, and so forth—toward their policy objectives. This has the effect of fomenting more action by mobilizing coalitions of the willing from the broader population of actors.

While orchestration is widespread across many areas of world politics, the climate regime is perhaps unique in the extent to which policy entrepreneurs engaged in "webcraft" to orchestrate new actors to complement intergovernmental diplomacy.[65] Transnational governance of climate change had been building over the course of the regime but took a large step forward around the creation of the 2015 Paris Agreement.[66] In the lead-up to Paris, the UN secretary-general, the UNFCCC secretariat, and the Peruvian and French hosts of Conference of the Parties (COP) 20 and COP 21, respectively, took a much more purposeful approach, mobilizing dozens of initiatives that ultimately came to include over ten thousand actors. At Paris, this orchestration function was then institutionalized in the UNFCCC process with the creation of the role of high-level climate champion, an individual each COP host appoints for a two-year term with the explicit mandate to spur first movers among cities, businesses, and other actors.[67] The following year, this function was further institutionalized into the Marrakech Partnership for Global Climate Action. Such mechanisms seek to mobilize a wider and wider mass of first movers.

Third, catalytic institutions promote iteration. Even though flexible commitments capture only what actors are willing to do, with increasing returns, what actors are willing to do changes over time. Catalytic institutions therefore create an ongoing process to record updates in actors' preferences over time. For example, by requiring new commitments every five years, the Paris Agreement allows increasing returns to be translated into new, stronger actions, which may then generate their own effects. A one-off commitment model would lack this ability to harvest the beneficial effects of past actions on subsequent actions over time.

Beyond Paris, iteration of commitment-making is a common feature of international institutions. The global trade regime has developed through progressive trade rounds, and arms control agreements have been iterative. The "framework and protocol" approach to global common issues like the ozone regime typically involves a series of increasingly stringent negotiated agreements, as was originally pursued in the climate regime.[68] While these latter agreements use collective agreements enforced with sanctions, they also exhibit catalytic elements in the sense that past cooperation alters future preferences. In trade, for example, the expansion of multinational companies enabled by early trade rounds reshaped the domestic politics of economic openness in major economies by creating powerful new constituencies for integration.[69] Paris replicates this logic but with individual (for both countries and other actors), as opposed to negotiated, commitments, which allows it to capture updates in actors' preferences more quickly and easily.

Finally and perhaps most important, catalytic institutions seek to reinforce and amplify the various mechanisms that drive path dependence, increasing the effect of prior action on subsequent action. As discussed above, many of the material, informational, normative, and political economy mechanisms through which increasing returns accrue fall outside the realm of political institutions. For example, reductions in the price of clean technologies are transmitted through markets and social norms diffuse through a wide range of processes.[70] However, institutions can play a number of complementary roles to augment increasing returns in each of the four types of causal processes described above—material, informational, normative, and political economy. Again, climate policy provides rich examples.

One, as discussed further in chapter 4, institutions can make material transfers to alter actors' preferences and capacity to act, what in climate or development policy is often termed capacity-building—training programs, experiential learning, technical assistance, and so forth. Donor countries often deprioritize these efforts because they are seen as "soft" interventions without tangible impacts that can be reported back to constituents. But a catalytic lens highlights their potential.

Many material transfers are simply negotiated bargains, but they become catalytic when, above and beyond providing a direct side payment, they increase the capability of the recipient to take action in the future. For example, a grant that helps pay for a renewable energy project reduces the one-off cost of acting. In contrast, a training and support package that improves a country's ability to run effective tendering processes for energy procurement could permanently reduce the cost of acting and strengthen the country's preference for decarbonization. Similarly, technical support that gives a country a better sense of what impacts it will suffer from climate change may increase the value it places on mitigation efforts. Although the Paris Agreement did not contain any firm commitments to increase funding for developing countries (arguably the Kyoto Protocol was superior in this regard) and contributions to the Green Climate Fund have proven small, it did create a Paris Committee on Capacity Building that oversees and supports developing countries' ability to formulate and implement national climate policies. Thus far, however, this process has yet to actually generate resource flows that have a catalytic effect.

Two, institutions can create processes for learning. For example, many international institutions have review processes or procedures to exchange best practices.[71] As discussed above, such information is catalytic when it creates demonstration and learning effects that boost actors' ability and willingness to act. In the Paris Agreement, review occurs through a collective Global Stocktake every five years as well as more limited review of individual NDC implementation. These mechanisms are fairly high level, so it remains to be seen how much actual learning comes from them.

Potentially more promising, a major goal of linking sub- and nonstate actors to the intergovernmental regime is to promote learning. The number and diversity of sub- and nonstate actors makes them excellent laboratories for climate policy, and many transnational initiatives explicitly promote information exchange.[72] The Paris system tries to enhance the epistemic benefits of sub- and nonstate climate action by creating structures to review and extract lessons from nonstate actors including an online platform to record their efforts, the climate action events at COPs, an annual Yearbook of Climate Action, and engaging them directly in the Global Stocktake. Still, as the legal scholar Kenneth Abbott notes, there is potential for the UNFCCC and other actors to play a more active role in enhancing the catalytic effect of these transnational elements of the regime.[73]

Three, institutions are key for enabling the normative effects of goal-setting and benchmarking, discussed further in chapter 5. Formal goal-setting can create a legitimate normative focal point around which actors can converge. This mechanism is unlikely to sway actors who do not wish to cooperate, but it can enhance efficiency and enable benchmarking among actors guided by a logic of appropriateness, as first movers are likely to be, while also empowering pro-action interest groups.[74] To the extent governments or other actors are sensitive about reputational critiques, explicit goal-setting raises the costs of noncompliance, and gives pro-compliance groups a hook for their arguments.[75] Indeed, once a normative goal has been set, it becomes possible to review actors' progress toward that goal. Benchmarking systems—which grade states against some standard and compare their compliance to others—are now common tools in world politics.[76] Although some advocates proposed creating a grading system for NDCs in the Paris Agreement as part of the enhanced transparency framework, countries balked at exposing their "nationally determined" contributions to the collective judgment of their peers or others (suggesting the power of such ranking systems). Still, the Paris transparency framework requires countries to report on their progress toward their NDCs and thus creates the conditions under which third parties can compare and rank national ambition.[77]

Four, institutions can also be catalytic where they allow constituencies for action on a long problem to grow. By actively supporting cities, businesses, investors, and other such actors to take action on climate change, the Paris architecture supports constituencies that can push for greater climate action at the domestic level. These efforts can have catalytic effects by influencing the process of national preference formation. The more sub- and nonstate actors in a country take ambitious climate action, the more they increase the likelihood of both their peers and the national government adopting strong climate policies as well.[78]

In sum, catalytic institutions

- set goals,
- mobilize first movers and incremental action through flexibility,
- iterate commitments actions, and
- increase the effect of prior action on subsequent action via
  - providing material transfers to alter future preferences and capacities,
  - creating processes for learning and information exchange,
  - enabling normative goal-setting and benchmarking, and
  - building constituencies for action over time.

Assessed with this theoretical lens, the Paris Agreement looks potentially stronger than conventional perspectives that see collective agreements with hard enforcement as the only way to overcome collective action problems.[79] It also widens the mechanisms through which iteration and flexibility can work beyond the technological and informational processes emphasized by experimentalist governance. To the extent the Paris Agreement contributes through catalytic mechanisms to shifting actors' underlying preferences and the larger problem structure of climate change, it may actually help create the conditions under which "harder" institutions become possible or even partially obviate the need for them. At minimum, the Paris Agreement offers a better approach to muddling through than what came before it because it directly tackles the early action paradox.

4

The Long View

ADDRESSING SHADOW INTERESTS

Two hundred years ago, the conservative British philosopher Edmund Burke famously described society as a contract between the dead, the living, and those yet to be. But have any of us actually signed our names to this contract? Future people will be profoundly affected by how we address (or fail to address) long problems but have no ability to affect us. Moreover, we can only dimly, imprecisely understand what those interests are. Without the ability to advocate for themselves, future generations can only depend on our charity or hope that our interests align with theirs. Of course, the same applies to us. The paradox is that although we are free to create the future as we see fit, so were our predecessors. We are just as much prisoners of their choices as our successors will be trapped by ours. Can we break this chain? If so, how?

The dilemma of shadow interests—the lack of agency of those affected by long problems—saps political power from the long term. This chapter asks under what conditions might an actor wish to act in the long-term interest and under what conditions that actor might gain power in the present. This political focus filters the many proposals and projects for intergenerational fairness through the lens of critical realism. It also helps look beyond the narrowly institutional focus of much of the work on this topic, considering not just how changes in the decision-making rules or the organization of government can promote long-termism but also what political strategies can shift the underlying balance of power and interests toward long-term outcomes.

Several types of interventions can help, at least to some extent, to overcome the problem of shadow interests. Representatives can aim to speak for the interests of future generations, or at least approximations of them, seeking to address the fundamental lack of agency that lies at the core of the shadow interest dilemma. However, it is then up to political actors in the present to act or not on that information. More potently, political actors can empower trustee institutions—courts, central banks, or even purpose-built future generation guardians—to act on behalf of shadow interests. Finally, some strategies, rules, and processes, such as capacity-building or participatory deliberation, are able to actually shift present actors' preferences toward longer time horizons.

The chapter focuses on the extent to which we can or cannot get around the problem of shadow interests and how to do so, not whether it is right to do so. I assume that in many cases, it will be beneficial to do so, given the dilemma of long problems, but I largely leave aside the knotty ethical questions around representing future generations, such as how they can authorize their present representatives and hold them accountable or how to address the plurality of potentially competing future interests.[1] Instead, the chapter asks a social scientific question: If we want to give our shadows force, how might we do it?

Representation: Voices for Future Generations

The previous chapter discussed how making information available and salient can help address the early action paradox by increasing the likelihood and ability of actors in the present to work in ways that support long-term interests. But what about creating actual representatives for future people tasked with assessing their interests and speaking on their behalf? Governments often create special commission(er)s, ombudsmen, or other kinds of advisory bodies for specific long problems, especially those of a scientific or technical nature like climate change, infrastructure development, or pandemic preparedness. More rarely and recently, governments are creating various kinds of representatives for future generations in general. Such institutions can be distinguished from the simple procedural requirement to consider information about long-term consequences (discussed in the previous chapter) in that they bring a new actor into the political process charged with actively intervening in political discussions about long problems. In other words, such institutions create a new source of political agency for the long term and thus go further than the informational mechanisms discussed in the previous chapter even if they work through similar ways of addressing uncertainty and low salience. If the challenge of shadow interests is future people's utter lack of agency, creating future-oriented institutional actors gives future generations an avatar in the present.

Representatives of future interests can not only make information about the future available and salient, they can also advocate and hope to persuade actors in the present to pay attention. In this way, institutions for representation can help link actors across time just as transnational networks can bridge actors across space. A repressed minority group in an authoritarian state may lack any means of directly challenging the dominant domestic rulers. But by building transnational linkages to allies abroad, who then lobby their own governments to pressure the oppressive regime on behalf of the minority group, even weak interests can gain significant political leverage through a "boomerang effect."[2] Analogously, while future people cannot actively forge such linkages, representative institutions can help mobilize potential allies in the present.

Such institutions vary enormously in their constitutional position, statutory basis, political independence, breadth of policy focus, formal roles and powers, composition, and other dimensions.[3] Some act as policy entrepreneurs, others as neutral brokers.[4] The New Zealand Parliamentary Commissioner for the Environment, discussed in the previous chapter, enjoys significant autonomy and a wide mandate to not just assess the environmental impacts of different measures but also to bring attention to new topics and to respond to citizen complaints. The CBO, in contrast, has endured as a powerful, autonomous actor even as Congress has cut back on the power of other such bodies, in part because it has scrupulously limited itself to the carefully proscribed functions its legal mandate requires.[5] It assesses; it does not advocate.

In the climate realm, the most influential body creating information about future outcomes, though it does not explicitly aim to represent future interests, is the IPCC. Over its thirty-year history, the IPCC has become increasingly salient. Its reports and pronouncements are widely covered in the news media and, indeed, are of such journalistic interest that they are regularly leaked. This prominence derives from the IPCC's ability to summarize the global state of scientific knowledge, with thousands of authors aggregating and adjudicating information from tens of thousands of scientific articles. It is further reinforced by the IPCC's status as a multilateral organization, with every government in the world ultimately signing off on its reports. It is difficult to imagine an institutional design that could create a more authoritative voice on climate science.

As argued in the previous chapter, the information a salient, authoritative institution like the IPCC provides can have real political weight. It can alter the balance of power and interests around future outcomes by mobilizing aligned interests and by setting the political agenda. Here, however, it is important to see how the agency the IPCC possesses enhances this function. Perhaps the best example of this ability was the way in which the IPCC's 2018 Special Report on Global Warming of 1.5°C altered how the climate problem was understood for a wide range of actors. At COP 21 in Paris in 2015, activists and vulnerable countries pushed hard for the goal of total decarbonization of the world economy, or net-zero emissions, by the middle of the century, with the aim of limiting overall temperature change to 1.5°C. Against significant odds—given the opposition of large emitters—they partially succeeded, securing a commitment to achieve a "balance between sources and sinks" of GHGs in the second half of the twenty-first century, in the effort to limit temperature change to 1.5°C, or at least "well below" 2°C.

Tied to this effort, the same pro-climate interests were able to ask the IPCC to deliver a special report on the new temperature goal. Although their role was purely technocratic, the IPCC authors (climate scientists) actively promoted a flurry of studies to flesh out the knowledge base on the difference between 1.5°C and 2°C. The new political goal reoriented the questions scientists were asking in a way that then shifted how the technocratic body came to define the climate problem.[6]

The resulting publication, pulled together in rapid succession by the global scientific community, was a blockbuster. After the report was released, media discussion of climate change surged by over 40 percent.[7] Essentially, the report changed the political understanding of the climate problem in two ways. First, it laid out how much worse climate impacts would be at 2°C compared to 1.5°C. Looking across oceans, agriculture, extreme weather, heat, and other vectors through which climate change disrupts human and natural systems, the report made clear the high stakes and value of achieving the more ambitious temperature target. It showed that the slogan vulnerable countries had begun to use, "1.5 to stay alive," was literally true for certain places and populations. Second, the report showed that achieving this goal would require a radical acceleration in mitigation, reducing global carbon emissions by 45 percent below 2010 levels by 2030, and achieving net zero by 2050. The media focused in particular on the 2030 goal, framing the report's findings as "12 years to save the world." This simplistic message was critiqued as misleading by climate scientists, including the report's authors, though some experimental evidence showed the deadline framing bolstered support for government action on climate change.[8]

It is of course difficult to tease out the specific impact of a single report, but it is notable how the document was picked up and frequently cited in two major developments following its publication. First, a number of governments, cities, states, regions, and businesses that were already taking climate action quickly reformulated their goals and objectives to align to the new understanding. Many of these actors began to adopt net-zero targets, promising to end their contributions to climate change in line with, or ahead of, the pathways laid out in the report. Just a few years after the report was published, over 90 percent of global GDP was covered by a national net-zero target.[9] Second, the report helped inspire an unprecedented wave of climate activism, including a potent youth movement, as the final section of this chapter discusses. For example, following the report's launch, Swedish youth climate activist Greta Thunberg often cited it and referred to its findings. In this way, even a technocratic institution like the IPCC was able, through its agency, to provide highly salient and authoritative information in a way that helped set the political agenda and mobilize political action in the present.

While the IPCC is the best-known example of such a body in the climate realm, an increasing number of jurisdictions are creating bodies that provide information and assessment of climate change at the national or subnational levels. Such bodies can be found across most European countries and also in places like China, South Africa, Chile, or the state of New York. Whereas some of these bodies are quite weak, providing only summaries of information in response to policymakers' needs, others have significant analytic resources and a mandate to proactively intervene in the climate policy process.

The UK Climate Change Committee (UKCCC) is an early and important example of the political weight such institutions can have. Like the examples cited above, the UKCCC has autonomy and authority as an expert organization charged with overseeing the implementation of the 2010 Climate Change Act. However, it also possesses a significant degree of agency, as its terms of reference empower it to put forward regular carbon budgets for the United Kingdom on its road to net zero. These budgets are technocratic analyses of what level of GHGs are permissible for the United Kingdom to remain on course toward its climate targets and include assessments of what pathways the government could pursue to get there. The UKCCC's mandate also requires it to assess the adequacy of the government's progress on implementation. This power to put forward specific targets, to which the government is statutorily required to respond, and to assess progress gives the UKCCC a clearer institutional role than many peer institutions in other countries, which are often limited to advice and review.[10]

In its first decade of operations, the UKCCC has become a significant political force for the long-term interest.[11] As with the above examples, the UKCCC's role is purely informational. But it has been effective in utilizing its agency to ensure the information it provides matters in at least two ways. First, it has a key agenda-setting function. The regular carbon budgets it puts out define the ambition that government policy is meant to reach toward and highlight a number of the specific policy tools and choices that the government could take up. A review found significant, though certainly not wholesale, adoption of the UKCCC's recommended carbon budgets, even over opposition from the Treasury and other powerful actors in British politics.[12] Tellingly, the UKCCC's longer term budgets were accepted while recommendations to accelerate emissions in the nearer term were not. Still, the UKCCC's agenda-setting power puts the onus on the government to reject and justify departures from its recommendations.

Second, the UKCCC's analysis now attracts significant political attention. The UKCCC was invoked hundreds of times in parliamentary debates in its first decade, five times more than the IPCC.[13] It is invoked by all political parties and by parties in government and in opposition, though more by opposition parties and more by left-leaning parties. In the debates on carbon budgets, two-thirds of all comments related to government policy explicitly invoked the UKCCC to urge greater action.

The IPCC and the UKCCC are examples of institutions that provide climate-specific information about the future, addressing, albeit only partially, the challenge of shadow interests. But governments have sometimes created broader institutions that aim to fulfill a similar role for the future overall. Across countries, we see numerous examples of commissioners, committees, ombudsmen, representatives, and even ministers for the future. These positions or bodies are tasked with reviewing the future implications of *all* policy decisions and bringing the envisioned concerns of future generations into present debates. Their ambition is to directly confront the dilemma of shadow interests.

Examples include Finland's parliamentary Committee for the Future, the Welsh Commissioner for Future Generations, Tunisia's Commission for Sustainable Development and the Rights of Future Generations, the Hungarian Parliamentary Commissioner for Future Generations, the Knesset Commission for Future Generations in Israel, Sweden's Council on Future, the Maltese Guardian of Future Generations, and the United Arab Emirate's Ministry of Cabinet Affairs and the Future. Such bodies vary enormously in the heft of the individuals who lead them, the formal powers they possess, their organizational capacities and budgets, and other features. On average, though, their role is primarily informational and agenda-setting. As the former Welsh Future Generations commissioner has stated, "I can't force anybody to do anything, and I can't force government to stop doing something."[14]

That limitation of course holds for the IPCC and other climate-specific institutions as well.[15] But in comparison to bodies like the IPCC or UKCCC, the track record to date for generic future institutions holds relatively few cases of clear successes. Again, the problem is the need for shadow interests to find willing allies in the present.

The Israeli example is revealing. Created in 2001, the Knesset Commission on Future Generations had more formal powers than any of the general purpose future bodies yet created. As an organ of the legislature, it had the power not just to advise but also to draft and propose new laws across a wide range of policy domains. Although the commissioner himself (a respected judge) could not vote, his ability to propose new laws gave future generations a direct agent in the present. The commissioner could also delay proposed legislation by querying its impact on the future, a potentially very powerful quasi-veto given the importance of timing in legislative processes.[16]

Despite these formal powers, it is difficult to identify where the commission had impact on legislative outcomes. A proposed constitutionalization of the principle of sustainable development was rejected by the Knesset. Ideas around introducing various procedural requirements to check the future impact of new laws were partially adopted but only resulted in "tick box" changes. None of the proposed laws put forward by the commission were adopted, and the quasi-veto was never deployed. Most damningly, the Commission on Future Generations endured only five years. When the commissioner stepped down in 2006, he was not replaced, and the underlying law was itself undone in 2010. Ironically, a number of other future-oriented bodies have similarly short life spans, such as the Hungarian Parliamentary Commissioner for Future Generations, whose powers were later subsumed into a broader human rights body, or the UK Sustainable Development Commission, which lasted only ten years.

Boston's analysis shows how a lack of political support enervated the Israeli experiment in long-term governance. The Knesset commission did not emerge from a party manifesto or the demands of interest groups or social movements. Rather, it was the brainchild of a single influential politician. When that legislator retired, there was no political basis to maintain its support. As Boston writes, the problem of shadow interests bites hard: "Ultimately, building and sustaining the required political support is critical. Self-evidently, such support cannot come from the people who are supposed to be represented by such a body (i.e. future generations)."[17]

Moreover, the generalist focus of the Israeli example and its cognates highlights the importance of connecting future interests to concrete allies in the present. As Boston argues, any institution with a broad, open-ended, future-oriented mandate is likely to find it harder to mobilize support from civil society groups than a comparable institution with a narrower remit (e.g. to defend a healthy environment), an identifiable group (e.g. indigenous peoples) or a set of rights (e.g. the rights of children). This is because few politically influential lobby groups will regard a generalist, future-oriented institution as a crucial vehicle for advancing their particular interests and concerns. In short, they will lack a sense of "ownership" of the institution in question. As a result, any commission (or other public entity) for future generations runs the risk of having few friends and defenders.[18]

As proposals for various future-oriented commissioners, ombudsmen, and other such bodies proliferate, it is critical to take seriously the political challenges of shadow interests.[19] Future-oriented institutions need to be endowed with sufficient political weight to stand against other political actors in the present. The next section considers how such weight may come from constitutional rules that give future-oriented institutions legal authority. But mobilizing sufficient political support to create this kind of constitutional power is of course itself undermined by the challenge of shadow interests. It is perhaps for this reason that generic future-oriented institutions have not often endured, even though they may be normatively desirable compared to single-issue bodies in that they could more holistically consider various trade-offs and priorities.[20]

Instead, future representatives tied to specific constituencies and political interests in the present, like budgets or climate change, seem to have more weight than "generic" future institutions, perhaps because they can more easily find allies in the present. This gives them a basis in politics that "the future" in the abstract can likely never have. After all, there may be some actors in the present willing to ally themselves with future generations purely for the sake of solidarity with people who do not yet exist, but such temporal altruists are not likely to hold much political weight in the present.

This problem highlights a general contradiction facing institutions that aim to speak to future interests or on behalf of future generations. Their impact is often tied to technical, impartial authority they possess or mobilize, typically rooted in scientific information. Bodies like the IPCC or the UKCCC are designed to rise above the fray of political contestation. Although their outputs may be (and are) used by political actors, their role would arguably be undermined if they appeared overly partisan, or at least that is the conventional wisdom. This creates an inherent limitation on representatives of future interests because it means they cannot compete for power and influence in the present like an activist, an interest group, or a political party would.

Given this problem, it is perhaps worth reconsidering whether it makes sense for bodies like the IPCC or the UKCCC to remain scrupulously "above politics." Many scholars have noted the inevitably political nature of "boundary institutions," which straddle science, politics, and policy, including those that seek to assess the future, no matter how they are staffed or structured.[21] Most observers would now agree that remaining outside politics is infeasible and so advocate for procedural processes like transparency and citizen and stakeholder consultation to manage the limits of technocracy without giving up the fundamental neutrality of the technocratic body.

If we take the challenge of shadow interests seriously, we must accept that such institutions play political roles, and should think about their design accordingly, not apart from the political system but embedded within it. For example, could we imagine future-oriented bodies commenting on proposed legislation or even evaluating different political parties' manifestos? In the climate realm, there is significant debate among scientists in how vocal they should be, with many deciding to be more active in the political process—for example, as seen through the March for Science that began in 2017. It seems plausible that as long problems grow, giving technocratic bodies an increasingly important role in how we define political problems and solutions, such bodies will need to expand their role in the political process.

In sum, representatives of future generations can alter the politics of the present through providing information and raising salience, akin to the informational mechanisms explored in the prior chapter. What differentiates them, however, is their agency to advocate for and persuade, not just inform. Through their agency, representatives of future generations can at least partially overcome the dilemma of shadow interests.

Trusteeship: Insulation and Tying Our Hands

If representatives for future interests cannot mobilize sufficient allies in the present, what about giving real decision-making power to actors oriented toward the long term? At root, the problem of shadow interests is the inability of future people to affect politics and policy in the present. An obvious solution would then be for governments to create bodies with both the interest and the power to act for the long-term benefit. They could mandate new institutions to advance future interests, insulate them from short-term pressures, and critically, grant them real power.

When political actors delegate power to others who (are meant to) share and implement their interests, they create agents. Instead, when leaders wish to check their own impulses, they create trustees, actors with different incentives and the ability to act independently—for example, courts, special investigators, central banks, international human rights commissions, and so forth. Trustees are one way for governments both to insulate policy from short-term pressures and to tie the hands of decision-makers who might seek to work against the long-term interest. Representatives can bring information about future outcomes into the policymaking process, mobilizing potential future-oriented actors in the present and helping them set the policy agenda. Trusteeship goes one step further by giving future-oriented actors explicit decision-making power.

Of course, the problem of shadow interests cannot be fully solved through delegation. At the heart of trusteeship is the perennial dilemma of institutionalizing credible commitments. First, a decision-maker in the present must be sufficiently motivated toward the longer term to create a trustee. Second, they must remain bound by the trustee, even when they have incentives to do otherwise. Because an authority that can give power away may also be able to take it back, trustees' ability to push against short-term interests faces an inherent limit. Even the strongest institution can be undermined or overruled if enough political opposition to it emerges. The challenge is therefore to find formal and informal ways to institutionalize trustees' power so as to make it more difficult for their decisions to be reversed or for their authority to be revoked.

Although delegation to trustees is perhaps the most potent way to insulate decision-making or tie one's hands, other institutional arrangements can go at least some distance toward these goals, intentionally or not. Chapter 2 noted how different types of political institutions—regimes types, term limits, electoral and party systems, and so forth—put decisions under more or less short-term pressure. It is important to note, however, that insulating decision-makers does not alter their fundamental preferences. It just means that they have more scope to resist short-term pressures if they want to. For decision-makers who seek to act in the long-term, insulation will help them do so. For those who do not seek to act in the long term, insulation will matter little. Incentives may change but not preferences. In this way, insulating decision-makers in the present is similar to giving them better information about future outcomes. It helps those already motivated to act in the longer term to do so but cannot necessarily make them want to act in the longer term interest. Without delegation to a trustee, the decision-maker remains unchanged.

For this reason, addressing long problems by delegating to bureaucracies that are tightly accountable to political leaders, like regulatory agencies, are imperfect solutions. Delegation to an agent creates a degree of insulation but does not reshape preferences. In the United States, for example, regulatory rule-making has become an important domain for advancing climate policy in the face of legislative gridlock in Congress. The 1970 Clean Air Act gives the Environmental Protection Agency (EPA) significant authority to set standards around air quality, which it has used to restrict carbon pollution. The EPA's specific regulatory decision-making is insulated from Congress, though it is constrained by administrative law provisions and the courts and its authority could be revoked or overruled by Congress in the future. Most importantly, however, as an executive agency, the EPA's ability to do so is strongly conditioned by the political interests of who is in the White House. When pro-climate administrations wish to use executive authority to advance climate goals, they can do so. But such moves can be, and have been, reversed by subsequent anti-climate leaders.

Similarly, political leaders often attempt to tie their own hands without actually giving up their power, using weaker forms of so-called commitment devices.[22] Again, fiscal policy is an instructive area in which leaders often introduce pledges, laws, procedural requirements, or sometimes even constitutional rules that put boundaries around public spending. One clear example is the European Union's (EU's) 1992 Maastricht Treaty, which created the fundamental rules for the European currency by limiting member states' fiscal deficits to 3 percent of GDP and overall debt to 60 percent of GDP. As a European treaty, the rules are legally binding on all Eurozone states, representing one of the most robust institutional commitment devices policymakers can deploy. In practice, however, Eurozone countries have followed the rules to varying degrees. The original rules allowed limited exceptions when countries face recessions, but these were then expanded over time as various crises created a short-term interest for greater flexibility. Moreover, countries that broke the rules often avoided sanction, particularly if they were powerful member states who could pressure European institutions into not acting against them. And over time, many experts have come to see the original rules as overly strict, encouraging countries to reform them as political and expert support for fiscal rigidity has decreased.[23] In sum, the commitments enshrined in European law in 1992 proved malleable in practice. If those who wished to flout the rules have enough power to avoid punishment, or if the consequence of following the rules is too unappealing, or simply if ideas on how best to address a problem change, then commitment devices prove unable to constrain behavior.

Self-constraint is always challenging for governments because of the time-inconsistency problem. But for long problems, the dilemma of shadow interests further limits self-control strategies in two ways. First and most obviously, actors in the present constrain themselves to advance their own, present interests, which may or may not overlap with those of future generations. Fiscal spending limits provide a case in point. For governments with poor credit, credible constraints on future spending may be needed to access loans in the present to fund current governmental priorities, not necessarily to invest for the benefit of the future. Moreover, a strong commitment to pay back loans often reflects the political power of present creditors—by definition, holders of capital—which is often greater than those who may benefit from additional spending, such as poorer populations reliant on welfare. To help address long problems, decision-makers in the present have to be motivated to constrain themselves to achieve a future outcome.

Second, self-control is temporally unidirectional. A current actor can at least partially constrain a future actor via institutional fetters or other tools, but a future actor lacks all agency to constrain those of us living in the present. For a long problem in which achieving benefits in the future requires making potentially costly changes in the present, such as climate change, self-constraint offers little practical value. The challenge is not limiting our successors but rather creating more options for them by limiting ourselves. While we can attempt to constrain our future selves, we can never reach back in time and constrain our prior selves.

For the above reasons, neither the de facto insulation from short-term pressures that some political institutions provide to current decision-makers nor the efforts by present actors to control themselves can do very much to address the problems caused by shadow interests. Instead, creating new actors, trustees, who are designed to hold long-term preferences, who are insulated from short-term pressures, and who hold real decision-making power, offers a stronger tool.

To create a trustee, it is necessary to shape or select the preferences of the trustee to diverge from the near-term interests of current leaders and to ensure the trustee has real decision-making authority even when objections inevitably arise.

To orient a trustee's preferences toward the long term, selection, tenure, capacity, and professional norms are key. A trustee may be nominated by a political leader, but if they are too aligned to that leader's preferences, their independence may be compromised. Using an independent expert panel to nominate a shortlist of candidates is one way to avoid such a difficulty. By the same token, if a trustee must rely on a leader for reappointment, their incentive to act against that leader's interests is weak. Therefore, a long term of office, perhaps without reappointment, can enhance independence. Resources also matter. If a trustee must depend on decision-makers to hire staff or access information, it will find itself having to make deals with those it may need to stand up to in order to perform its job. Finally and perhaps most important, trustees need the right, mission-oriented organizational culture. To the extent they are technocratic experts, trustees are accountable to the norms and expectations of their profession or epistemic community. A central banker who crashes the economy will face shame and opprobrium from her peers, as will a judge whose legal reasoning is self-evidently weak or spurious. Because technocrats, by definition, typically base their identities around their area of knowledge and expertise and because their future career prospects depend on their prestige within their profession, organizational culture can powerfully shape their preferences.

Once the right trustee is in place, it must be able to maintain its authority in the face of resistance from short-term interests. Typically, trustees rely on institutional authority for this. For example, central banks have the responsibility to set interest rates; no other actor can pretend to do so. In practice, though, institutional authority sits in dynamic tension with political interests. Should sufficient opposition mobilize against a trustee, its independence could be compromised or abolished. Institutions are harder to get rid of when they are protected by a larger number of veto points in a political decision-making system. For example, if the trustee serves merely at the pleasure of the executive, they can change when the leader or the leader's preferences change. When they are created through a legislature, abolishment may require a majority vote. And when embedded in the constitution, a supermajority may be needed to stymie a trustee.

Maintaining trustees' authority is about more than institutional checks, however. Trustees understand that their institutional authority is predicated on political agreement. Should they push too hard against the interests of political authorities, they can expect their power to be curtailed. For example, in many countries, constitutional courts enjoy strong formal and informal autonomy from political leaders, including the ability to carry out judicial review of legislation.[24] However, even the most autonomous courts must remain sensitive to how other elements of the political system perceive their actions. Even if they believe they are following sound legal arguments, judges are, on average, often loath to break too much from political leaders, lest they provoke a backlash that would undermine their power. Famously, for example, US president Franklin

Roosevelt threatened to add new justices to the Supreme Court after the conservative majority blocked a number of key provisions under his New Deal. Although Roosevelt was unable to carry through on his threat, the court did shift to a more favorable stance, allowing the New Deal to proceed. Just as legislators create laws with an eye toward how they will withstand judicial review, so too do courts act with an eye to how the political autonomy of their institution can be maintained.[25] Even though there is thus a constant tension between trustees and political leaders, the balance is not static. Trustees may accumulate power over time, as constitutional courts have done in most established democracies, or they may be brought in check by institutional reforms.

Of course, the hardest challenge is imbuing trustees with real decision-making power in the first place. The dilemma of shadow interests is recursive; if you do not have an incentive to act in the longer term interest, why would you have an incentive to give some of your decision-making power to a trustee to act in the longer term interest? We therefore need to examine the political conditions under which delegation to trustees becomes possible.

First, decision-makers in the present must actually have an incentive to tie their hands. Consider the creation of independent central banks as trustees tasked with monetary policy. While the functions and governance of central banks has evolved over time, since the 1960s, the trend has been toward greater independence. In wealthy countries, the 1970s saw a resurgence of inflation that economists believed required credible commitments to sustained higher interest rates to solve. In particular holders of capital, a politically powerful group, worried inflation would negatively affect them. In response, governments further empowered their central banks to act independently to curb inflation. Interestingly, looking across all countries shows that prior inflationary episodes, not other kinds of financial shocks, are correlated with increases in central bank independence.[26] Wealthy states then exported this lesson to the developing world through the international financial institutions. Poorer countries that received assistance from the International Monetary Fund (IMF) or World Bank agreed to adopt a range of economic reforms, and central bank independence was often at the top of the list.[27] Again, a cross-national survey shows that receiving support from the IMF is a strong predictor of increased central bank independence.[28] The point is that the political choices that made contemporary central banks independent emerged in response to the immediate incentives political leaders and powerful interest groups faced in the 1970s, 1980s, and 1990s. Politicians tied their hands not because of great foresight but because of immediate needs and pressures.

Second, to have the confidence to delegate to a trustee, a decision-maker must not feel that the trustee will soon act contrary to their interests. Ironically, here the shortness of a decision-maker's time horizon may actually work in favor of the longer term. As long as the trustee can be expected to constrain behavior only sometime in the far future, it is more likely that they will be granted power in the present.

The creation of human rights bodies perhaps shows the logic of this condition. Such bodies, like the Inter-American Court of Human Rights or European Court of Human Rights saw their memberships swell following the waves of democratization in the 1970s and 1980s. Why would governments consent to put themselves under the scrutiny of such institutions? Newly established democratic regimes were eager to signal their human rights credentials and differentiate themselves from the authoritarian communist or military regimes that had preceded them and so had an immediate incentive to show that they were credible. But they also wanted to prevent backsliding by locking in liberal institutions that might constrain their successors.[29] Delegating authority to international human rights bodies presented both a strong signal of the new regime's intentions and a credible attempt to constrain future leaders who might be less than democratic. It was easy for the newly democratic regimes to sign up to human rights constraints in part because they never imagined it would limit their scope for action.

These two conditions suggest that delegation to trustees is likely to occur in temporally bound windows of opportunity when decision-makers win power who can gain some benefit from delegating and when they do not expect delegation to cost them very much. Such delegation will typically only be possible when some constituency is actively pushing for it and happens to gain power. As noted in the previous section, this makes it very unlikely we will see delegation to generic "future-oriented" trustees. Rather, we can expect trustees to emerge around specific topics of political importance, like monetary policy or human rights. Because trustees are only likely to be established in these narrow windows of opportunity, we expect them to be undersupplied on average.

Perhaps the strongest, though still imperfect, example of a climate policy trustee-like institution is the California Air Resources Board (CARB). Signed into law by then-governor Ronald Reagan in 1967, CARB is mandated with cleaning up California's air. It has broad authority to make rules regulating any aspect of air pollution, including GHG emissions, reaching many parts of the economy. For example, CARB both creates the rules governing the California Cap-and-Trade Program and runs the market on a day-to-day basis. It also sets highly influential vehicle manufacturing standards. The legislature and governor set out CARB's broad goals, such as California's current target of climate neutrality by 2045 or its interim target of achieving 60 percent renewable energy by 2030, giving CARB a long-term orientation. Under this guidance, CARB is in charge of creating the plans and rules for how this will be achieved and indeed operates many of the programs that implement its plans. For example, California's landmark 2018 climate law, SB 100, is only thirteen pages long. An earlier bill, SB 32, which included the requirement to reduce GHGs by 40 percent by 2030, was only two pages. The details are left largely to CARB, which sets rules following a period of comment and consultation in accordance with administrative law procedures, and which is ultimately accountable to the courts.[30]

Although CARB operates as a regulatory agency, it enjoys a relatively high degree of both de jure and de facto independence that has given it trustee-like qualities. It is governed not by a single political authority but by a sixteen-member board appointed to six-year terms by the governor and the legislature, with seats on the board reserved for certain areas of expertise (e.g., auto emissions) and perspectives (e.g., environmental justice). This structure certainly does not make it immune from interest group pressure (indeed, its governance model gives regulatees formal channels of influence), but these have not gridlocked the institution. It has a large, expert staff funded by revenue obtained not by the legislature but from its own regulatory activities, such as the auction of emissions permits. More informally but perhaps just as importantly, CARB's leadership has been very effective at proactively establishing its authority by maintaining strong relationships with political leaders.[31] It has over time built up a reputation for competence, backed by California's success in improving air quality and lowering emissions, which has arguably encouraged political leaders to increase its discretion.

Critically, CARB's authority is conditional on the strong environmental consensus that has pervaded California politics.[32] Beset by smog in the middle of the twentieth century, Californian voters strongly favored environmental action and so welcomed the creation of a new regulatory agency in the 1960s, when CARB was set up.[33] Since that time, the state's population has become even more committed to environmental goals, following the emergence of the modern environmental movement. Although California is a large consumer of energy and produces oil, it does not have a significant auto industry or other interest groups powerful enough to block action.[34] Indeed, some of its most politically potent economic interests, such as farmers, are extremely vulnerable to drought or fires. Across decades, political leaders of different parties have therefore seen delegation to CARB as a useful way to achieve the policy goals their constituents desire, especially given the myriad distributional conflicts that emerge from the trade-offs around how climate and energy policy can be operationalized.[35] Although CARB has certainly had to navigate the shifting political interests of the legislature, it has never faced political leadership fundamentally opposed to its work.

What would it take for more climate trustee institutions to emerge in other jurisdictions? Beyond CARB, we already see less authoritative examples like the IPCC or the UKCCC and similar committees, discussed above, where climate-specific actors have been created to provide information and to advise. Although this role gives them agenda-setting power, there are as yet no cases where such bodies have been granted full policymaking discretion, judicial authority, or other such powers. CARB is notably unique and of course exists under very favorable political conditions. It is possible that the more advisory bodies could be developed into full-fledged trustees if decision-makers face sufficient political pressure to lock in their commitments to climate goals and fear their successors may seek to roll back from the ambitious targets they have set. For example, what if lawmakers empowered an expert committee to set an economy-wide carbon price based on the best available science? In the United States, the use of social cost of carbon in regulatory standard setting takes a step in this direction, although this remains under the control of political leaders. One could imagine that over time, relatively independent climate committees could accrete more power through political strategies that establish precedents and build constituencies that back them, as courts and central banks have done.

However, those other cases may not represent feasible models for the climate challenge because of problem length. Human rights institutions emerged only after the trauma of twentieth-century wars and genocides, and independent courts took decades, even centuries, to carve out authority independent of political leaders. For a long problem like climate change, trustees instead need to be established preemptively; a decades-long historical process will not generate a solution in time. Independent central banks, for their part, have benefited from having powerful interest groups in the present that push for them. As climate change proceeds and generates more existential politics, it may generate analogous conditions that could see the emergence of new trustee institutions. But these other examples suggest that the establishment of such institutions, and their accumulation of authority, will likely play out over a longer timescale. In this sense, long-term-oriented trustee institutions may be an *outcome* of climate change, rather than a solution that can prevent it. I return to this idea in the conclusion.

Because creating new trustees for the long term is difficult and likely to emerge only over time, a more feasible strategy may be to instead increase the future orientation of the most powerful *current* set of trustees in many, though not all, political systems—courts of law. Indeed, this process is already underway.

At present, a wave of climate litigation is rapidly advancing through the courts of many jurisdictions. The number of climate-related suits doubled from 2015 to 2021.[36] Some are brought against companies and some against governments. The legal issues they raise are similarly heterogenous, ranging from company directors' fiduciary duty to consider climate impacts, to the loss and damage climate change resulting from previous emissions has already caused, to disputed claims over greenwashing. The Dutch courts have been particularly fertile grounds for these arguments. In *Urgenda Foundation v. State of the Netherlands*, the court ordered the Dutch government to increase the ambition of its climate target. In *Milieudefensie et al. v. Royal Dutch Shell plc.*, the court imposed a similar mandate on the oil major, ordering it to reduce the emissions associated with the products it sells sooner rather than later.

Within this heterogenous array of litigation, a number of cases have argued that the rights of future generations, or of youth today, require the protection and intervention of the courts, seeking to confront the problem of shadow interests. Similarly, a number have invoked the precautionary principle, which similarly requires actions in the present to reduce potential future harms. In Germany, a youth group (notably, not future people) successfully sued the government to enhance its near-term emissions targets, with the constitutional court arguing it was unfair to delay such measures into the future. In Colombia, another group of youth successfully petitioned the supreme court to order the government to formulate and implement steps to address deforestation in the Amazon. However, in other high-profile cases like *Juliana v. the United States* where young people have argued that government inaction on climate change is harming their future, courts have declined to intervene. In the Dutch cases mentioned above, while the rights of future generations or youth today are mentioned in passing, the primary legal reasoning follows from the current obligations of governments and companies under environmental and human rights laws. The concerns of future people are only obliquely referenced in most cases.[37]

For a long problem like climate change, it is important to see the development of these kinds of legal norms as an evolutionary process. Legal principles emerge through a mixture of case law, strategic legal activism (which drove the cases cited above), and the changing understandings of jurists about what the law is and what it requires. For climate change, this process is likely to continue to evolve as climate impacts deepen. Although we cannot predict how and whether strong legal protections for future generations will be adopted, other experiences hint at how they might develop. Over the second half of the twenty-first century, activists and jurists drew on a growing body of international human rights law to find new ways to protect the rights of people who were previously thought to lack standing in courts. For example, in the United States, an eighteenth-century law, the Alien Tort Claims Act, was repurposed to allow anyone in the world to bring claims against multinational corporations that had done them harm, giving people around the world protection under international treaties in US federal courts. Similarly, courts developed a doctrine of "universal jurisdiction" over grave crimes, with a Spanish court famously bringing charges against former Chilean dictator Augusto Pinochet for crimes committed in Chile, even though he was physically in London. In these ways, a combination of human rights treaties, activist lawyers, and changing ideas around legal norms created a set of protections for people who previously had no recourse to courts.

Could future people win similar protections? Constitutional scholars and legal philosophers have proposed just that through ideas of intergenerational justice. The German philosopher Jörg Tremmel writes, "The concept of 'checks and balances' . . . was designed to protect part of the population against the 'tyranny of the majority.' Today we need temporal checks and balances in order to protect future generations from a 'tyranny of the present.'"[38] In other words, transforming courts into trustees for the future can help overcome the problem of shadow interests.

This form of "legal long-termism"[39] will require expanding the number of international and domestic legal provisions that speak to the concerns of future people and also continuing strategic litigation from plaintiffs aiming to translate these norms into decisions and precedents. New legal provisions can take many forms. They can, for example, provide for fundamental rights (e.g., the right to a healthy environment) or enshrine general public policy provisions (e.g., the requirement to maintain fiscal discipline) as overarching long-term goals that courts must act to defend.[40] As noted in the introduction, such ideas are already very common in national constitutions (over 41 percent refer to future generations).[41] A survey of over five hundred law professors found that jurists supported protections for future nearly as much as for current people.[42] Advocating for stronger and more explicit protections for future people in international and domestic law can bolster the transformation of courts into trustees for the future. For example, in 2021, the UN secretary-general's landmark report *Our Common Agenda* proposed a new "Declaration on Future Generations" in which states could further establish such norms.

Of course, empowering judges to serve as trustees for future generations also faces important limits that derive from the challenge of shadow interests. After all, judges are still people in the present. They may be insulated from short-term pressures to varying degrees, but they cannot be fully immune to them. Moreover, even if they decide the law requires protections for future interests, they are of course required to serve the interests of the present as well.

Perhaps more fundamentally, legal cases tend to require plaintiffs. Because they are merely shadow interests, future people cannot file lawsuits. They instead rely on proxies—legal activists, youth groups, and so forth—to act on their behalf. As with the informational mechanisms discussed above, this means that future interests are always refracted through the understandings and preferences of actors in the present. But in the legal context, it also introduces questions of who has standing to speak on behalf of future generations, especially if those future generations may have varied interests.

Most centrally, the strength of courts on checking political decisions varies widely around the world and over time. Courts are, in many places, strong trustees. But in plenty of jurisdictions, they are weak, and in no place can it be assumed that they are fully immune from pushback from other elements of the political system. Should courts go too far in protecting future interests, they could see their authority revoked or curtailed. We can therefore expect these tools to be most effective when judicial independence is strongly protected in institutions and where opposition to action on climate change is not so strong or universal that an anti-climate coalition could emerge strong enough to undermine judicial authority. This will be the case in many but not all jurisdictions. Notably, many petrostates lack strong independent judiciaries, perhaps because the centralization of power linked to resource rents discourages the development of institutional checks. This tool may therefore only have limited traction in some of the jurisdictions most central to the climate challenge.

As these considerations show, empowering trustees necessarily comes at a cost to the power and discretion other actors in the political system. Trustees do not "work" unless they impose real constraints. Even though this adjustment of power might be efficient and legitimate to address a long problem, it is important to consider both sides of this trade-off. Few people would want to live in a society in which urgent, immediate needs cannot be addressed because powerful trustees overweigh long-term objectives. The arguments in this book show why, politically, such situations are unlikely to emerge (there will be strong political pressure to brush long-term interests aside), but to the extent we are successful in creating more long-term-oriented trustees, these political frictions will grow. As with courts or central banks, it will be important to see climate trustees are part of a wider political system that combines institutions that derive legitimacy from participation or representation (e.g., legislatures) with those that are instead delegated to act by legitimate authorities (e.g., bureaucracies or courts).[43] As the next section explores, long problems necessitate not just trustees but also more participatory institutions like climate assemblies.

Despite these limits, building stronger legal protections for the future represents one of the most promising practical ways to put real political weight behind future interests. The question thus arises, how can this approach be advanced? As with international human rights law, new international treaties and constitutional provisions can help change legal norms over time.[44] A corresponding international legal obligation, perhaps in a new legal instrument on the rights of future generations, could reinforce this norm. Importantly, advocates of such measures should see this project as not just legal or philosophical but also profoundly political in that it reshuffles power from present interests to future ones. This will ultimately require fundamental changes in social norms to have traction. The instruments of modern human rights law emerged from the unique political conditions that followed the horror of state violence in the twentieth century. The impacts of climate change, as they increasingly batter our societies throughout the twenty-first century and beyond, may provoke a similar consensus of never again, creating political conditions that could empower jurists to act in the interests of future generations.

Horizon-Shifting: Changing Preferences

In the short term, preferences are relatively fixed. But over the course of a long problem, they can vary. One way to overcome the challenges of shadow interests, the lack of agency of future people, is therefore to extend the time horizons of actors in the present—to create conditions under which we value future interests more. Of the three strategies presented in this chapter, this is the slowest and likely the hardest. But it is also the most powerful because it actually builds political power in the present around longer-term outcomes, confronting the problem of shadow interests directly.[45]

At first blush, seeking to shape or even engineer preferences may seem Orwellian in the context of contemporary politics, particularly in democratic societies. Certainly, telling people what to think is normatively undesirable and perhaps unlikely to alter true preferences in a meaningful way. However, there are a number of political institutions and strategies that can induce both material and normative changes in preferences by changing the rules of the game or via deliberation and persuasion. The large normative literature on intergenerational justice and long-termism makes a strong case for why we should, as individuals and as societies, adopt a longer time horizon.[46] Scholars of global environmental governance have advocated for strategies that make sustainability a fundamental norm for navigating the Anthropocene.[47] This section reviews some promising examples for how we might do so.

Consider material preferences first. Perhaps the most powerful way to build material incentives for the long term is to alter the time frame on which material value is measured and apportioned. Many observers have criticized companies for responding more to short-term pressures than long-term imperatives, particularly when driven to show positive quarterly earnings reports that keep stock shares high. CEOs and investors themselves have extolled the need to build companies oriented toward the longer term, developing principles and frameworks for doing so through leading business forums. Researchers have also found evidence that companies' long-term investments are systematically undervalued by their shareholders[48] and that firms with longer time horizons perform better overall.[49] In response, a wide range of advice and consulting services have arisen to help companies become more oriented toward the long term.

But of course, overcoming short-termism is not just a question of helping CEOs, boards, and investors devise ways to build better companies. As chapter 2 argued, it is a systemic challenge that requires, ultimately, fundamental shifts in public and private institutions to comprehensively reshape incentives, including a significant role for public regulation. For example, some reformers have called for changing securities law to make corporate earnings reports less frequent. But others have argued it is not the frequency of reporting per se that matters but rather the fact that market analysts publish predictions for firm performance and companies have a strong incentive to show investors that they are meeting those predictions.[50] The point is that piecemeal reform is unlikely to fundamentally alter systemic short-termism in how investors behave and how companies are managed. It must be attacked root and branch.

Proposals for how to do so abound. Companies could be required to link executive compensation to longer term outcomes, particularly for shares they may hold. Similarly, the fund managers who decide whether to invest in companies could be required to link compensation to longer term outcomes. Capital gains taxes could be weighted by the length of time a share is held, or an excise tax could be placed on short-term trades so that holding stocks for longer periods becomes increasingly attractive. Regulators seeking to build a longer term–oriented economy could bundle these and other reforms into a holistic package that change the incentives all actors face in order to influence the time horizon of the market overall.

Note that the reforms listed above focus on changing not *what* companies care about but for *how long* they care about it. To the extent companies' time horizons extend, we would expect them to have a greater interest in addressing longer problems, such as climate change. But reforming *what* companies care about can also have an important effect on long problems, and indeed there is an enormous push from governments, civil society, and within the market itself to embed environmental and social goals in corporate aims.[51] To the extent corporate aims are expanded to include longer problems like climate change or many other sustainability challenges, the effect of this expansion in scope will also be to extend time horizons.[52]

Proposals for and examples of embedding broader values in corporate governance abound. Benefit corporations (B Corps) put environmental and social performance equal to profit as part of a triple bottom line. The World Economic Forum has evangelized the idea of stakeholder capitalism in lieu of shareholder capitalism. Leading business schools are teaching future corporate managers how to build purpose beyond profit into the companies they will run.

To have impact, this movement will need to scale beyond promising experiments and positive ideas to a more fundamental rewiring of contemporary economies. For example, in the United Kingdom, campaigners for the Better Business Act are proposing to rewrite the basic law governing publicly listed companies.[53] The present law requires directors to prioritize financial interests above all other factors. The proposed language would alter this to include environmental and social goals as well, while also creating clearer reporting requirements against these broader objectives.

Analogously, reformers have proposed looking beyond narrow measures of the economy overall, principally GDP, to prioritize more holistic measures that include social and environmental welfare. For example, the Wellbeing Economy Alliance proposes prioritizing environmental performance, meeting human needs, equality, good governance, and other measures into the primary ranking of economic success and attainment.⁵⁴ Mainstreaming such measures would shift the system overall, advocates suggest, because so many actors' preferences—politicians' reelection goals, relative power between countries, people's assessment of the overall state of affairs—are expressed in terms of GDP.

From a political perspective, however, it is important to ask, is the problem that actors' preferences are influenced by the wrong measures or is it instead that the wrong preferences are determining what we measure? Put another way, alternative measures to GDP already exist, so why are they not being taken up and used more often? Inertia is likely part of the explanation, but a political economy lens would also consider how such alternative measures would come at a cost to current powerful interests. From this perspective, the question is not so much what measure to use but rather how to build political power around the kinds of preferences, including long-term preferences to which better measures would align.

Given the radical departure from current systems that rewriting corporate law or replacing GDP with more holistic measures would represent, it is perhaps difficult to imagine large economies fundamentally embracing such deep reforms to the economic system in the near term. However, for long problems, deep changes can accumulate over time. Fascinatingly, an increasing number of countries are adopting regulations to require companies to disclose their climate data and to publish transition plans or otherwise weaving net zero requirements into the broader fabric of economic and financial regulation. These changes are being driven by pro-climate activists and particularly investors. To the extent they succeed, they will, by altering what companies and investors care about, also extend the time over which they care. In other words, climate politics may be one of the most viable near-term paths to inject more long-termism into the market overall.

To the extent markets are reshaped to prioritize longer-term outcomes, the results could be truly profound for the political economy around long problems. Future interests would be more matched with the preferences of powerful economic actors in the present. The dilemma of shadow interests—the fundamental lack of agency of people in the future—would diminish as actors in the present took on their cause. Political support to create trustee-type institutions would increase. Information provision tools would be able to mobilize not just activists but large segments of the economy around the interests of future people. Even though the problem of shadow interests would certainly not disappear, the horizon of political and economic attention would extend.

Of course, preferences are made not just from material factors but also ideational ones: changing beliefs, norms, and worldviews.[55] Again, the development of the human rights regime is instructive. Over more than a century, many people around the world came to believe fundamentally different things about the rights and duties we owe to each other. Of course, such changes did not lead to a utopia in which all rights are always upheld. But they did fundamentally alter political institutions and behavior in profound ways. Can we imagine a similar shift in temporal preferences? Philosophers have proposed as much,[56] and there are many political strategies that can build support for longer term interests, such as leadership, education, and activism—all of which can currently observe working to change preferences in the realm of climate politics.

By leadership I mean the scholar and university president Nannerl O. Keohane's idea of articulating a common vision for a group and bringing people to that vision.[57] Elites of different kinds can have a powerful influence on setting the terms of political discourse, which can shape individual beliefs and even identities in powerful ways. We see this kind of normative leadership in the climate realm.

For example, the institutions and actors described above as representatives of or trustees for future generations like the IPCC, climate change committees, or jurists can influence norms as well as information or the specific policy areas over which they have authority. Indeed, such actors can be policy entrepreneurs and persuasive advocates as well.[58] Even though the IPCC's mandate is strictly to provide scientific information, the practical effect of its increasingly dire reports is to serve as a kind of clarion call for increasing action. As noted above, the 2018 1.5°C report was perhaps particularly influential in reshaping how many political actors and individuals conceptualized the climate problem.

Beyond these expert bodies, other actors can also help change preferences by invoking the long term. Political leaders can adopt rhetoric that mobilizes their constituents' longer term interests, and cultural figures can popularize such beliefs.[59] These kinds of elite signals have typically played a key role in long-term processes of social change.

For example, UNFCCC executive secretary Christiana Figueres, who steered the Paris Agreement to adoption, cites hope as a core political strategy behind the successful negotiations.[60] Although many doubted that an agreement could be adopted following previous failures, Figueres understood her role as projecting a sense of optimism that she hoped would be self-fulfilling. Indeed, it was. After her term, Figueres went on to found, together with her adviser Tom Rivett-Carnac, an organization called Global Optimism that seeks to keep alive—indeed, to weaponize—the idea that addressing climate change is feasible. After all, governing for the future "rests upon a deep and abiding hope that humanity has a future, that this future will be worthwhile, and that even the most challenging problems will not prove overwhelming nor the worst disasters unmanageable."[61] Doom, in contrast, can lead individuals to give up trying. Hope therefore has a key role to play in shifting time horizons to enable long-term governance.

Elites can influence norms through leadership, but institutions can change norms through society as a whole via capacity-building and education. In the international climate regime, dozens of bilateral and multilateral programs seek to help countries gain the technical expertise and capability to address the climate problem. The need for capacity-building and education has been recognized since the 1992 framework convention, and various programs have sought to generate attention and resources to the issue. Capacity-building is the most commonly identified barrier to climate action in countries' NDCs.[62] For this reason, the Paris Agreement created a Paris Committee on Capacity Building as a kind of clearinghouse for these efforts. It helps countries assess their capacity needs and aims to source support from both peers and donors. Total climate-related capacity-building development assistance currently stands at around $10 billion per year, nearly half of all climate-related development assistance.[63] However, the total amount is miniscule compared to the need.[64]

From a long-term perspective, however, the value of capacity-building is even greater. As chapter 3 argues, it could be one of the most catalytic areas of the Paris Agreement because of its potential to transform states' preferences over time.[65] In the policy process, a state's capacity to plan for and implement climate policies is intimately connected to what it deems both possible and desirable to achieve. The development of a cohort of officials charged with climate policy within a government also creates an internal constituency for climate policies within the bureaucracy. Both of these mechanisms can influence the ambition of policies that political leaders set by setting the range of what is feasible and what survives the policy process. Research has also shown that state capacity helps countries avoid conflict and other security breakdowns when natural disasters strike, bolstering resilience.[66] Capacity-building could therefore be a critical tool for reshaping climate politics over the longer term, though donor governments too often see it not as a strategic asset but rather a concession they must make in international negotiations.

Beyond government officials and processes, the international climate regime also includes provisions related to education of the population at large.

Article 6 of the 1992 framework convention pledges to promote educational and public awareness programs on climate change and its effects, a commitment echoed again in Article 10 of the 1997 Kyoto Protocol and Article 12 of the 2015 Paris Agreement. As with capacity-building, there is little in the international process that operationalizes these claims. Other UN bodies, such as UNESCO, the UN Environment Programme, and the World Meteorological Organization, have climate education campaigns, but these are relatively small in scale. Giving citizens accurate knowledge to navigate the changing climate in which we live could substantially alter states' and other actors' preferences over climate cooperation.[67] But countries have only superficially used the international regime to make investments in climate change as a tool to change what might be possible in the future.

Still, over the last decades, climate change has become a widely taught subject around the world. Today, climate change features in some way in the curricula of nearly all educational systems around the world from primary to tertiary levels, though the depth and precision of the content varies significantly.[68] Reviews have found that much climate education remains quite conventional, limited to the science of the challenge and awareness of some of its impacts and how they might be prevented or adapted to. Scholars have instead pointed to the value of more participatory and affect-driven approaches that teach not just the facts of climate change but build critical-thinking skills to understand how it might affect one's own society and how that society might best respond.[69]

This kind of mass education is perhaps the most powerful tool available to states for shaping norms, beliefs, and identities. Histories of the development of modern states point to the critical role of mass education in creating a common sense of identity.[70] But schools and other formal institutions are not the only way to shape norms and beliefs through education. One of the most fascinating developments in climate politics in recent years has been the emergence of climate assemblies—local or, in some cases, national groups of citizens who gather to deliberate on how to address climate change in their community or society.[71] Climate assemblies vary in their form, but all share the goal of creating a deliberative process through which ordinary citizens grapple with the problem of addressing climate change and put forward ideas for how their city, region, or country might address it. Some are formally established by political leaders, like the French Citizens Convention on Climate. Others emerge from bottom-up grassroots movements, like the Germany Citizen's Assembly on Climate. Some focus on a relatively specific policy challenge, such as reaching net-zero emissions, while others have an open-ended mandate to address climate issues writ large. Some make specific recommendations that are then taken up by political leaders, while others are merely advisory or hortatory.

To date, hundreds of assemblies have emerged at different scales. While examples can be found around the world, the majority have taken place in Europe, with salient national examples including the French and German examples cited above, but also Denmark's Climate Assembly, Climate Assembly UK, and the Irish Citizens' Assembly, and examples in Austria, Spain, and elsewhere. Alongside these national exercises, smaller scale deliberative assemblies on climate-related issues can be found around the world, ranging from flood prevention in Uganda, to waste disposal in Brazil, to recovery from extreme storms in the Philippines.[72]

Interestingly, developing countries have pioneered the use of more deliberative techniques in the UNFCCC negotiations. As host of COP 17 in Durban in 2011, South Africa faced the unenviable task of getting agreement on a way forward following the breakdown of trust at COP 15 in Copenhagen in 2009. To break the gridlock, the COP presidency invited negotiators to an Indaba, a deliberative practice in Zulu and Xhosa communities, which the official invite described as "establishing a common mind or a common story that all participants can take with them. In successful Indabas, participants come with open minds motivated by the spirit of the common good, listening to each other to find compromises that will benefit the community as a whole."[73] In the end, COP 17 agreed on a negotiating program that ultimately led to the adoption of the Paris Agreement in 2015. Similarly, Fiji, as host of COP 23 in 2017, instituted a Talanoa Dialogue, the practice of using storytelling to reach understanding and consensus in Pacific communities.

Citizens' movements as well as political scientists and philosophers have long been interested in the ways in which intentional public deliberation among a microcosm of a polity can improve public policy. Such "deliberative mini publics" are posited to improve democratic decision-making by, for example, surfacing informed public opinion, getting social buy-in for complex trade-offs, and giving potentially underrepresented interests voice. They provide little space for narrow interest groups to advance policies that benefit themselves at the expense of others. For these reasons, advocates of deliberative democracy see it as a vital complement to thin conceptualizations of democracy limited to electoral competition for power.[74]

Deliberative assemblies also have an important ability to shape norms and beliefs, including around longer term outcomes, particularly when combined with the foresight practices described in chapter 3.[75] Theorists of deliberative democracy note how assemblies can encourage participants to listen to opposing viewpoints and seek consensus. They create conditions under which people can genuinely persuade others or be persuaded themselves, using what Habermas terms the "forceless force of the better argument." They can also allow for the detailed consideration and integration of technocratic and scientific information as well as other kinds of knowledge like community norms or practices, which can be particularly useful for a topic like climate change.

Although deliberation can have various positive qualities, the most relevant point for long-termism is how it may lengthen participants' time horizons or lead to them to give greater attention and value to shadow interests. As we might expect, research has shown that individuals who participate in climate assemblies tend to develop greater concern regarding the topic and shift their policy preferences toward greater action.[76] What is interesting is one of the reasons for this shift seems to be the development of a greater concern for future outcomes and interests, perhaps through the mobilization of "type 2" or "slow thinking."[77] Deliberation allows individuals to better understand and empathize with future generations and therefore understand better what their own preferences on the well-being of future people might be.[78] Such considerations rarely come to the fore in "normal" political discourse, and so the special environment of a climate assembly creates the right assemblage of information and affective inputs to allow it to emerge.[79] In this way, climate assemblies may extend the time horizons of ordinary citizens, though more systematic research on this potential is needed.[80] Similarly, many foresight techniques require participants to actively imagine and construct future scenarios, bringing the potential reality of a future path to the front of their mind. The act of thinking through and imagining possibilities can help to reshape what an individual or organization values or not.

To be sure, citizens' committees often have limited power to directly influence decision-making. However, research has also found that such bodies can indirectly influence the political process as well. For example, researchers have found that people are more likely to support even costly climate policies that emerge from citizen assemblies compared to those enacted through conventional legislative processes.[81] In places where climate assemblies have taken place, participants considered their biggest impact to be adding momentum and legitimacy to political decisions to act more aggressively on climate change by showing broad public support.[82]

Thus far, most climate assemblies have been ad hoc or one-off affairs. But as societies face increasing long problems, we may benefit from integrating such deliberative tools as a normal part of our political systems. For example, the Irish Citizens' Assembly, though not a permanent constitutional organ, has met on an ongoing basis over several years to consider a number of significant challenges the nation faces beyond climate change. Just as the dilemma of overcoming shadow interests may require greater reliance on trustee institutions, so too it might lead over time to more reliance on instruments of deliberative democracy.

Finally, regardless of the institutions that shape politics in a society, political activism and civic engagement is perhaps the primary way to shift preferences because it can alter both what people want as well as the power structures that aggregate preferences into policies.[83] On the individual level, activist groups and social movements work to persuade their fellow citizens of a certain course of action, aiming to generate support for certain policy goals across a society. Activist groups then seek to mobilize individuals to exert pressure on political leaders, aiming to shift the preferences of the government and therefore to alter policy. This pressure can work through various channels. For example, mobilization gives political leaders information about what citizens really care about. When people take time to write a letter or join a protest, it demonstrates a high-level commitment to an issue. The political scientist Erica Chenoweth estimates that fundamental political impact results when as few as 3.5 percent of people in a society join a social movement, as this both signals and accelerates a broader social change beyond activists themselves.[84] To the extent political leaders wish to (at least appear to) be responsive to citizens' preferences, activism gives them important information about how to do so.

Perhaps more significantly, mobilization can also build and shift political power. For example, activists who can mobilize enough political support can engage in agenda-setting, forcing their preferred issue into the political discourse and requiring others to react. They can also bring greater salience to a topic, shifting power away from insiders by creating more majoritarian politics, as discussed above.[85] Finally, normatively, activists may simply persuade political leaders over time that their preferred policy is the right one.

Climate-focused activism dates from the origins of the international regime but has in more recent years undergone a step change in scale and impact. "Insider" and "outsider" activist groups had been at loggerheads around the 2009 Copenhagen summit, but the 2015 Paris Agreement provided a focal point for the activist movement to coalesce around.[86] A broader coalition of social groups, including labor, feminists, and others, increasingly targeted climate meetings and aimed to add their voices.[87] And around the time of the

2018 IPCC report, new movements like Extinction Rebellion, the Sunrise Movement, and Fridays for Future reached a new level of scale and attention, mobilizing millions of people across dozens of countries.[88]

At the same time, we observe both public opinion polls and even, for the first time, some electoral decisions shifting in favor of greater action on climate change. Polling shows rising levels of concern over climate change, particularly in Europe, where activism has been strongest.[89] The 2019 European Parliament elections saw many voters declare climate change to be a top priority, with the Greens significantly increasing their seats after promoting a strong climate platform. In the United States, the 2020 Democratic Party primary saw candidates rapidly shift their positions toward more aggressive action on climate as it emerged as a top priority for Democratic voters.[90] Notably, there has been on average less climate activism to date in the Global South, though this may change as climate change's impacts and efforts to address it accelerate.

A key part of this mobilization has of course been the role of youth activism. The Fridays for Future movement and related efforts have mobilized millions of schoolchildren and young people in support of stronger action on climate change. For long problems, youth activism may be particularly significant. Any kind of activism, of course, emerges from the interests of people in the present. Shadow interests cannot take to the streets. Still, we might expect younger people, on average, to have longer time horizons than older ones, essentially by definition. Certainly, preferences over climate policy show a stark generational divide.[91] But it is not clear how much these patterns generalize. Young people can be concerned about their futures, but equally they can myopically focus on the present. Similarly, older people can be quite invested in the future, perhaps particularly when concerned about the prospects for their descendants. Youth activism on climate change has been significant, but young people have been less mobilized on, say, ensuring pensions or social welfare continue to pay benefits long into the future, even though they would benefit from such outcomes. It is difficult to identify a systematic tendency for youth activism to prioritize the interests of future people across all issues, although it clearly does in the realm of climate change.

This cross-issue variation repeats a pattern observed above vis-à-vis expert commissions and special commissioners. General purpose "future-oriented" actors seem to lack political heft compared to those focused more narrowly on issues for which there is at least some constituency in the present. Young people may care more about climate change, but it is not clear their time horizons are systematically longer across the board. To the extent this is true, it carries implications for efforts to institutionalize youth representatives in political systems as proxies for future generations. Many proposals for longer term governance include versions of this idea—for example, adding youth representatives to governmental bodies.[92] In the UN system, youth is an officially recognized constituency group. Representing youth in governmental bodies is worth doing simply because they represent a large portion of the current population and so merit political voice and power. But it is less clear to what extent young people are systematically more long-term-oriented than older people, even though certainly they are with respect to climate change. As these complexities show, shadow interests are fundamentally different from the interests of those in the present. Our political institutions require new ways of accounting for their complete lack of agency.

5

Endurance and Adaptability

ADDRESSING INSTITUTIONAL LAG

The Ise Jinju temple in Japan is both one of the oldest and one of the newest buildings in the country. First built over one thousand years ago, it has been rebuilt scores of times since. Every twenty years, the community meticulously disassembles and reassembles the structure, allowing the skills and knowledge of its construction to persist from generation to generation. In this extraordinary model of self-replication, linked to Shinto beliefs around death and renewal, lay important lessons for governing long problems.

Long problems create institutional lag. Once they are set up, institutions persist. But over the course of a long problem, institutions need to have a capacity to change to match evolving circumstances and needs. How can we devise laws, constitutions, or international organizations that, like the Ise Jinju temple, can both endure and renew?

Observers of politics have always seen institutional lag as a problem. Thomas Jefferson, in a letter to James Madison,[1] adopted the extreme position that institutions lasting more than one generation, which he put at nineteen years, were unjust:

> It may be proved that no society can make a perpetual constitution, or even a perpetual law. The earth belongs always to the living generation. . . . The constitution and the laws of their predecessors extinguished them, in their natural course, with those whose will gave them being. . . . Every constitution, then, and every law, naturally expires at the end of 19 years. If it be enforced longer, it is an act of force and not of right.

Madison responded by objecting to the obvious impracticality:

> Would not a Government ceasing of necessity at the end of a given term, unless prolonged by some Constitutional Act, previous to its expiration, be too subject to the casualty and consequences of an interregnum? ... Would not such a periodical revision engender pernicious factions that might not otherwise come into existence; and agitate the public mind more frequently and more violently than might be expedient? ... I can find no relief from such embarrassments but in the received doctrine that a tacit assent may be given to established Governments & laws, and that this assent is to be inferred from the omission of an express revocation.

The American founders were debating how to ensure the democratic legitimacy of their new country, but their discussion highlights a more general problem of how to institutionalize the right balance of change and continuity. In a world of only short problems, this dilemma would be muted. A problem would arise, society would devise some means to address it, and move on. There would be no need for rules to endure long enough to generate institutional lag. Only in this counterfactual world of the constant present could Jefferson claim, "the earth belongs always to the living generation. . . . They manage it then, and what proceeds from it, as they please, during their usufruct."

In practice, institutions and, increasingly, problems endure beyond a single generation. We cannot just create and uncreate institutions, even if it were practical to do so, because the problems they target span significantly into the future. But precisely for this reason, other aspects of a problem's structure have more scope to change because they have more time in which to do so. In this way, longer problems generate more risk of institutional lag. Chapter 2 noted how this dilemma has hampered multilateral governance of climate change—for example, by fixing the definition of industrialized countries at 1992 levels or by being slow to take seriously the need to adapt to climate change, not just to prevent it.

At heart, addressing institutional lag is a dilemma of combining long-term stability with a capacity for continuous updating, endurance with adaptability. The scholar of environmental politics Oran Young puts the dilemma succinctly: "A central challenge going forward in efforts to govern complex systems is to find ways to achieve nimbleness or agility in responding to nonlinear changes, abrupt transitions, and surprising developments without sacrificing the durability of governance systems needed to make subjects take them seriously as guides to behavior."[2] Put another way, the popular business book *Built to Last* quips, "To be built to last you must be built to change."[3]

How can we build institutions that help societies remain "steadfast yet flexible"?[4] What political conditions do they require, and what strategies can generate them? This chapter considers three tools. First, it explores how goal-setting can give institutions a degree of endurance that nonetheless allows for updating. Next, it considers how governance can be more reflexive, focusing in particular on the role of sunset and review clauses to create "forcing moments" for institutional updating. Third, it considers how trigger mechanisms and reserves can combine the two goals of continuity and adaptability. Throughout, the chapter considers how policymakers can best manage the ongoing tension between endurance and adaptability.

Before proceeding, it is helpful to introduce two concepts. First, we can distinguish first-order and second-order institutions. The former set the rules by which the rules are made; the latter are the rules that govern a certain problem. First-order institutions include national constitutions and intergovernmental organizations. Second-order institutions are laws, regulations, judicial decisions, treaties, soft law, or other governance rules. On average, we should expect first-order institutions to be more durable than second-order institutions. Interestingly, though, the average constitution lasts only seventeen years (a point to Jefferson?).[5]

Second, political systems differ in their capacity for change depending on how many "veto players"—actors whose assent is needed to change the status quo—there are.[6] Veto players can be formal, such as a legislative house or the presidency, which might both have to agree before passing a law. They can also be informal and circumstantial; for example, several parties might need to agree to muster a parliamentary majority, or a number of key interest groups might need to agree before political leaders will contemplate a certain course of action. Creating more veto players can enhance durability but also undermines adaptability, particularly in first-order institutions. If veto players can block changes in the rules around who is and is not a veto player, they are unlikely to be convinced to give up that power. Nearly all first-order institutions like national constitutions require broader agreement and more onerous procedures (e.g., supermajorities or special conventions) to amend than second-order institutions like laws. The UNFCCC is an extreme example. Because the UNFCCC is governed by consensus, each of the 196 member states is a potential veto player. This design makes change extremely difficult, burdening the global climate regime with significant institutional lag. More nuanced strategies for ensuring endurance are needed.

Endurance: Goal-Setting

If institutional lag occurs when institutions update slower than other elements of the problem structures they govern, it may seem counterintuitive to consider endurance as part of the solution. After all, for long problems, the challenge is typically too much stickiness in institutions, not too much adaptability.

In fact, some endurance is necessary to distinguish adaptation from continuous transitory reactions. Without a degree of continuity, institutions would become momentary reflections of the current state of other elements of the problem structure, not really institutions in a meaningful sense. Endurance is needed because the ability to make credible long-term commitments is a fundamental barrier to effective policymaking over time. If the rules of the game change easily, it will be impossible for governments to commit to, for example, long-term monetary policy, fiscal borrowing, preserving cultural heritage, conserving land for nature, large infrastructure projects, strategic research and development and industrial policy, or other policies that require sustained and consistent action over time. In the climate realm, policy durability is similarly critical. Actors must chart pathways to zero emissions over decades. They must make investments in adaptation whose benefits will accrete over time. And even after net zero is reached, there is likely to be a need to sustain large-scale carbon removal for decades or more, even as the impacts of climate change intensify.

Part of the challenge of long problems is therefore to create institutions that endure while minimizing the dangers of dysfunctional lock-in. This section therefore considers how institutional endurance can be achieved in a way that does not undermine the capacity for adaptation, which is in turn taken up by the following sections.

Goal-setting, mentioned in chapter 3 as a component of experimentalism and catalysts, is one way to combine endurance with malleability.[7] Political actors set goals for a wide range of issues: limiting inflation, sending objects or people into space, eliminating a disease, achieving a certain level of literacy or education across a population, developing new technologies, or reaching net-zero emissions.[8] High-level global goals have been set through the Millennium Development Goals (MDGs) and their successors, the SDGs, as well as in more specific areas like the Paris Agreement's mitigation target of limiting climate change to well below 2°C.

In Oran Young's compelling summary, goal setting

> seeks to steer behavior by (i) establishing priorities to be used in allocating both attention and scarce resources among competing objectives, (ii) galvanizing the efforts of those assigned to work toward attaining the goals associated with the resultant priorities, (iii) identifying targets and providing yardsticks or benchmarks to be used in tracking progress toward achieving goals, and (iv) combating the tendency for short-term desires and impulses to distract the attention or resources of those assigned to the work of goal attainment.[9]

In this way, goal-setting contrasts with rule-making as a governance strategy. Rule-making seeks to achieve certain outcomes by creating incentives for or against specified behaviors. Goal-setting instead aims to mobilize behaviors that push toward a desired outcome. The two are obviously complementary. For example, having set a goal of net-zero emissions, a government may adopt rules on car emissions standards to work toward that objective.

To provide this steering function, goals must have a degree of endurance. If actors expect priorities to shift in the short term, they have little incentive to seriously invest in efforts to achieve the stated goal at any point in time. But if instead, goals can credibly endure long into the future, they have the potential to anchor actors' expectations over time. How and under what conditions is this possible?

First, goals can be formally institutionalized. Official procedures and timelines for setting and revising goals give actors the ability to lock objectives in for a certain time. For example, China has made planning a central component of its policymaking process. While other features of China's authoritarian system can be observed around the world, such as single-party rule, China is perhaps unique in how heavily the governance system institutionalizes goal-setting.[10] Because China's vast policymaking apparatus requires coordinating different ministries, layers of government, and other state or quasi-state bodies, which often have divergent preferences, goal-setting provides a critical tool to generate policy coherence under conditions of "fragmented authoritarianism."[11]

In practice, this means setting an array of concrete targets and subtargets. The primary instrument through which top policymakers define core social, economic, and environmental objectives are five-year plans. Many of goals in the plans are quantified, with subtargets then allocated to specific ministries, subnational governments, firms, or other entities to deliver. For example, China's fourteenth five-year plan, issued in 2021, contained specific targets on GHG reductions, renewables deployment, energy efficiency, afforestation, and other key areas. These overall targets are then passed on to relevant ministries to decompose into specific sectoral plans and targets, which are then assigned to relevant bodies across the policy system for implementation. In this way, goals are not just statements that can be forgotten or shifted but the core organizing framework for policymaking.

Although the five-year plans focus on half decades, they also often refer to or enshrine many of the longer-term targets China has set. For example, in 2012, China pledged to eliminate absolute poverty by 2020, which it achieved. The fourteenth five-year plan went further and included a number of targets for 2035, the time frame for China to reach "socialist modernization" including in economic growth, technology development, international relations, and sustainability, not least China's longer term decarbonization targets.

While China represents an outlier in both the extent and degree of formal institutionalization with which it sets goals, many governments have institutionalized goals of some kind or another. Some are broad aspirational visions, such as Saudi Arabia's Vision 2030 plan or Egypt's Egypt 2030 plan. Others are concrete and quantitative, such as the numerous "missions" that the Indian government sets in areas ranging from rural electrification, to renewables deployment, to literacy or health outcomes. At the regional level, the African Union, on the fiftieth anniversary its predecessor, the Organization of Africa Unity, adopted Agenda 2063 as the continent's vision for the next half century.

Countries also set joint goals at the multilateral level, such as eliminating smallpox (set in 1967, achieved in 1980). Indeed, one of the most fascinating developments in global governance over the past decades is the way in which multilateral goal-setting has become formally institutionalized on a massive scale.[12] At the UN Millennium Summit in 2000, countries set in motion what became eight MDGs to be achieved by 2015. These goals focused on economic, health, social, and environmental topics measured by twenty-one indicators. Then, in 2016, countries followed a significantly more elaborate process to develop seventeen SDGs, each with several subtargets and dozens of indicators, which are updated over time. Unlike with the MDGs, countries review their progress toward the SDGs each year at a High-Level Political Forum on Sustainable Development. In this way, goal-setting has become an overarching institutional form in the multilateral system as a whole.

Perhaps one of the underappreciated aspects of the UN climate process is its ability to catalyze national goal-setting and planning processes. In addition to requiring regular NDCs, the Paris Agreement invites countries to submit "long-term low greenhouse gas emission development strategies." From 2010, the UNFCCC has supported countries to create national adaptation plans. Like the NDCs, these longer term strategies vary in detail, ambition, and authority. But in some countries, the multilateral nudge has helped create a process for goal-setting and forward planning that has enhanced governments' ability to do long-term governance. For example, Bangladesh's National Adaptation Plan, adopted in 2022 after significant consultation and technical inputs, lays out a three-decade plan for protecting one of the most vulnerable countries from the ravages of climate change. The plan lays out high-level goals and principles, as well as 113 specific interventions to be taken, all to be reviewed and updated every five years. Although implementing a plan of this ambition, which is estimated to cost $230 billion through 2050, would not be easy for any country, its existence likely gives Bangladesh a better chance than it would have otherwise. In contrast, the US states of Florida and Texas, also highly vulnerable to climate change, lack similar forward planning exercises on climate impacts.

Second, goals can credibly endure even in the absence of formal institutionalization when they have high political salience and electorates, interest groups, or other stakeholders actively punish or reward political leaders for achieving them or not. For example, US president John Kennedy pledged in 1961 to land an astronaut on the moon by the end of the decade. A goal's salience is of course in large part a function of how much the public or other stakeholders care about an issue and hold leaders to it, but salience is also partially influenced by political leaders as a tool to make their goals more credible. Political leaders seeking power often declare ambitious goals to attract support—for example, taming inflation, reunifying the country, or achieving a technological breakthrough. Many, perhaps most, such goals get forgotten. But when leaders make a big deal of them, they effectively invest some of their reputational capital into achieving the goal. By tying themselves to a target, they run the risk of being held accountable for meeting it. For this reason, the more they pin themselves to the goal, the more credible the goal becomes, all else equal, because the growing reputational cost of missing the goal increases the leader's incentive to deliver it. In this way, even noninstitutionalized goals can gain long-term credibility if they get enough salience.

Of course, the endurance of these kinds of political goals are bounded by the tenure of political leaders. If the leader is expected to leave or lose power in the medium term, the goals they personally tie themselves to may go as well, making them less credible. High-frequency political turnover therefore naturally makes this kind of political goal-setting more difficult. This may lead to the conclusion that there is an autocratic advantage in goal-setting, but the reality is more nuanced. Some political leaders in democracies remain in office over long periods, and goals may be associated with political parties as a whole or adopted more generally, so may last beyond a single leader. For example, Kennedy died just two years after his moonshot and never lived to see it achieved. At the same time, some autocratic leaders find themselves in highly unstable positions, frequently at risk of losing power, which can make their goals precarious.

In sum, goals can be made credible over the long term either through formal institutionalization or through political strategies that increase their salience and thus the reputational costs of failing to meet them. The latter may also sometimes evolve into the former. For example, Chinese president Xi Jinping announced a 2060 net-zero target in September 2020 at the UN General Assembly. A year later, this was adopted both into China's formal pledge under the Paris Agreement and into the domestic planning processes. At the international level, the substance of the MDGs drew heavily from an earlier report by the Organisation for Economic Co-operation and Development (OECD) outlining development needs for the twenty-first century. In this way, the exact content and form of a goal may evolve, but the process of goal-setting endures over time.

But even if they can endure, how can goals then shape actors' behavior? Again, both formal and informal models can be observed. Although the mechanisms through which goals drive action are not unique to long problems (they apply also to short-term goals), they show ways in which longer term objectives can under certain conditions influence short-term behavior.

First, goals can structure the formal incentives political actors face. Again, the Chinese system provides a clear, even exaggerated, example. Party officials at all levels are generally rotated from position to position every five or, more rarely, ten years. All officials are evaluated by the party's powerful Central Organization Department via a formal system that reviews both the general qualities of officials (e.g., their aptitude and political views) as well as the concrete outcomes they have produced and the extent to which they deliver on the objectives they have been assigned. Specific performance indicators vary across roles and levels and are updated over time to match the objective laid out in the planning process but include economic growth rates, social stability, energy efficiency targets, performance of state firms, pollution control, social welfare objectives, or dozens of other indicators. Notably, the heads of counties, cities, or provinces will be expected to show widespread achievement of the targets agreed in the planning process if they expect to avoid sanction or secure promotion. This is certainly not to say that promotion is purely meritocratic—political connections and loyalties matter as well, especially at higher levels—but Chinese officials take goals and planning seriously because they are held to them.

Interestingly, incentives tied to goal-setting have become increasingly used not just in the Chinese Community Party but also by the world's largest companies. Performance metrics became a core practice of management schools, consulting firms, and companies from the 1980s.[13] The public sector of many countries has also embraced this trend, sometimes under the label of new public management.[14] In the climate realm, a small but growing number of companies have actually linked managers' pay incentives to decarbonization goals.

But whereas some goals are linked to hard incentives, others are not. For example, neither the SDGs nor the Paris Agreement impose any penalties if the goals they put forward are not met. Nonetheless, such soft goals can shape actors' behavior through various channels.[15]

First, goals can guide the behavior of actors motivated by a logic of appropriateness, as chapter 3 noted with respect to catalytic institutions.[16] If actors accept a goal as morally correct and authoritative, they may seek to act in accordance with it, even if there is no material incentive or formal requirement for them to do so. A striking example is the widespread adoption of the goals of the Paris Agreement by cities, regions, businesses, investors, and other organizations around the world. Many sub- and nonstate actors have long sought to address climate change, setting an array of their own targets. What is striking is how, after countries agreed in 2015 to limit temperature change to "well below" 2°C and to pursue efforts to hold them to 1.5°C, the same targets became widely adopted by these other actors as well. The largest initiative through which businesses currently set climate targets, the Science Based Targets Initiative, adopted the same goals as the Paris Agreement, initially requiring businesses to adopt a target aligned to "well below 2°C" but then upgrading this to a 1.5°C-aligned target. The same pattern can be seen in the largest networks for subnational governments. C40, the global alliance of megacities, adopted 1.5°C as its guide star. The Under 2 Coalition, the main network of states, regions, and provinces, which takes its name from the original UNFCCC temperature goal, shifted its objective to also incorporate the 1.5°C target. Together, these networks and their peers capture a giant share of the global economy. When the globally agreed target shifted, they updated their plans accordingly.

Second, goals can enable benchmarking, again noted as part of the catalytic mechanisms in chapter 3. Under certain conditions benchmarking, attached to long-term goals or not, can exert pressure on actors toward goals even if they are not intrinsically motivated to pursue them.[17] There is now a proliferation of quantitative evaluations of state behavior and performance in areas like the ease of doing business, human trafficking, money laundering, or pandemic preparedness. Such indexes often shape the expectations and beliefs of investors, voters, or other stakeholders in ways that exert real pressure. For example, research has shown that countries listed as having weak money laundering policies end up paying higher loan rates because markets grade them as risky, and those rated lowly on a human trafficking index end up adopting strong criminal measures against the practice to improve their scores.[18] By establishing longer term outcomes and ways of measuring progress toward them, goal-setting provides a clear basis for benchmarking and allows actors to be compared to one another. Again, the Paris Agreement provides a useful example. Climate-minded investors have created scoring systems for firms that consider how aligned a certain business is to the temperature goals of the Paris Agreement. This in turn enables them to create a "temperature rating" for their overall financial portfolios and redirect assets accordingly.[19] At the time of writing, financial institutions with nearly $150 trillion in assets have committed to align to the goals of the Paris Agreement through the Glasgow Financial Alliance for Net Zero.[20] When backed by investors, the Paris goals shape the material incentives for corporations that may have never heard of the Paris Agreement and whose incentives do not follow a logic of appropriateness.

In these ways, even soft or aspirational goals can influence actors' behavior. The necessary conditions are either that the actors themselves are intrinsically motivated by the goals, perhaps because their preferences follow a logic of appropriateness, or that they are influenced by other actors for whom goal-adherence matters. As the above examples illustrate, in the climate realm, these scope conditions are relatively expansive.

A final characteristic of goal-setting as a governance tool is how it can allow a diverse set of actors to cohere around a single target. The adoption of the Paris Agreement goals by a wide range of actors provides a recent example, but we can observe similar uptake around the SDGS. By embedding themselves across a wider array of actors, goals can become more durable.[21] Notably, when the United States withdrew from the Paris Agreement under President Donald Trump, American states, cities, businesses, and other actors formed the We Are Still In coalition and pledged to maintain efforts toward the goals of the agreement. This widespread endorsement of Paris—the coalition covered more than 70 percent of US GDP—reinforced the credibility of the Paris goals because it meant that the world's second largest emitter had not substantively dropped the agreement even if the White House had.

In sum, goal-setting can be an effective way to create policy durability when it is formally institutionalized or when goals are sufficiently salient so political leaders face real costs for deviating from them. Furthermore, goals can steer behavior when they are linked to positive or negative sanctions either formally or when actors follow a logic of appropriateness, or when an actor's citizens, investors, customers, or other stakeholders push them to support a certain goal. Together, these mechanisms make goal-setting a significant instrument for long-term policymaking but, of course in many cases, still quite limited.

It is worth reviewing the evidence to date on how well goal-setting has worked, especially for longer term goals at the global level. On the concrete outcomes, evidence is decidedly mixed. The MDGs saw substantial progress in areas like poverty and hunger reduction, disease prevention, and primary education but less progress in areas like sanitation or environmental sustainability (see table 5.1).[22] The more interesting, and difficult to answer, question is around what difference, if any, the goals themselves made to these processes, given the many drivers of outcomes. One study found that most lower income countries and sub-Saharan Africa countries saw a positive acceleration in trends on poverty alleviation, hunger, and other areas following the adoption of the goals.[23] In middle-income countries, the absolute progress was higher, but the acceleration in the positive trends was also less pronounced. These findings are of course merely suggestive. To examine the effectiveness of the MDGs as a governance intervention, we need to understand instead how they changed actors' ideas, preferences, and behavior.[24] For example, did governments increase their activities around the policy outcomes specified? Did donors provide more or better development assistance? These questions require careful process-tracing research.

What can we infer from this past record for the current global development goals, the SDGs? At the time of writing, progress toward the 2030 goals has been significantly set back by interrelated crises. A 2022 UN report found that the COVID-19 pandemic reversed four years' worth of progress on poverty alleviation.[25] The war in Ukraine and associated inflation shocks have put a broad suite of other indicators, including especially food security, at real risk. More qualitatively, the SDGs have become more salient and institutionalized than their predecessors. The SDGs have been adopted into many national planning processes and nonstate actor commitments and have become a central

Table 5.1. Achievement of the Millennium Development Goals 1990–2015

Goal	Outcome
Goal 1: Eradicate extreme poverty and hunger	Percentage of world population living in extreme poverty fell from 47 percent to 14 percent from 1990 to 2015. Undernourished people fell from 23.3 percent in 1990–1992 to 12.9 percent in 2014–2016.
Goal 2: Achieve universal primary education	The primary school net enrollment rate in the developing regions has reached 91 percent in 2015, up from 83 percent in 2000.
Goal 3: Promote gender equality and empower women	Improvements in girls' education in some regions.
Goal 4: Reduce child mortality	The global under-five mortality rate has declined from ninety to forty-three deaths per one thousand live births between 1990 and 2015.
Goal 5: Improve maternal health	From 1990 to 2015, the maternal mortality ratio declined by 45 percent.
Goal 6: Combat HIV/AIDS, malaria, and other diseases	New HIV infections fell by approximately 40 percent between 2000 and 2013. The global malaria incidence rate has fallen by an estimated 37 percent and the mortality rate by 58 percent.
Goal 7: Ensure environmental sustainability	Averting ozone depletion, increasing clean drinking water access. Global emissions of carbon dioxide increased by over 50 percent in 1990–2015.
Goal 8: Develop a global partnership for development	Official development assistance from developed countries increased by 66 percent in real terms between 2000 and 2014, reaching $135.2 billion.

Source: United Nations 2015.

reference point for development. Less clear evidence has emerged that the SDGs are significantly affecting resource allocation, however. One comprehensive evaluation therefore concludes, "Overall, our assessment indicates that although there are some limited effects of the SDGs, they are not yet a transformative force in and of themselves."[26]

Alongside this mixed empirical record, it is important to recognize the potential pathologies that governance via goal-setting can create. For example, consistent with the idea of institutional lag, goals may be too rigid, driving dysfunctional behaviors. China's commitment to eliminating community transmission of COVID-19 provides a relevant example. Initially, the strict approach allowed China to remain almost entirely open domestically following the initial outbreak (though closed internationally) without suffering adverse health impacts. Countries as diverse as New Zealand or Thailand benefited from similar approaches. But as the virus grew more transmissible and other countries built immunity through vaccination, China became in 2022 the only country to continue pursuing an elimination strategy, necessitating continuous mass testing, lightning lockdowns, and severe international travel restrictions at the same time as other countries had gained a greater sense of normality. As the later sections in this chapter argue, adaptability and updating are key complements to goal-setting.

Goal-setting can also be highly problematic if progress toward goals is evaluated with inappropriate or narrow indicators that generate perverse outcomes. After all, goals often aim at final outcomes that are not necessarily decomposable into measurable indicators. Actors also invariably find ways to game the system by developing methods to score high on indicators but not necessarily to advance the underlying goal.[27]

Finally, goal-setting may run into problems when actors face multiple, potentially competing goals. If one of the purposes of goal-setting is to prioritize scarce resources and attention around critical priorities, too many goals can undermine their effectiveness or even generate conflicts. For example, observers of the SDGs have noted the potential points of friction between them—for example, preserving land for climate mitigation may undermine the food security of some populations. Without a system to integrate different goals and, where necessary, decide on priorities, the SDGs become less effective at steering behavior.[28]

Similarly, in the Chinese system, local officials face myriad competing demands, such as economic growth, environmental protection, social welfare provision, regime stability, and so forth. While political stability remains the foundational priority, local officials struggle with important trade-offs between, for example, economic growth and environmental protection, with the latter traditionally losing out to the former. Officials seek to prioritize those targets they believe their superiors will favor, but this then causes implementation gaps despite—indeed, sometimes because of—the comprehensive set of key performance indicators officials have to meet.[29]

Overall, goal-setting is a critical part of addressing long problems but needs to be paired with other tools to be effective. It can be a key part of addressing institutional lag when it provides continuity for the adaptive tools explored below.

Reflexive Governance: Updating Institutions

Because a greater problem length increases the risk of institutional lag, institutions addressing long problems need to be updatable. In theory, updating is always possible. As the legal theorist and sixteenth-century statesman Francis Bacon put it, "It is a perpetual law that no human or positive law can be perpetual"[30]—perhaps an inspiration for Jefferson in his debate with Madison. Legislatures that make one law can always make another. Countries can negotiate new treaties. Societies can write new constitutions. But as Madison emphasized, in practice, updating is hard. Creating new institutions takes time and effort, and interests that benefit from the status quo will fight to defend it. When they are veto players, they can block change. More subtly but equally powerfully, the status quo anchors actors' expectations. For these reasons, substantial changes require a political actor to expend time and energy, and they often must wait for circumstantial windows of opportunity to open. Updating is undersupplied.

However, various methods exist to make updating part of institutions' normal work. Together, we can term these approaches strategies for "reflexive governance."[31] Broadly, reflexive governance approaches involve, first, tools to combine knowledge across different disciplines and perspective. As in the foresight process described above, diverse perspectives are needed to better account for the wide array of variables and uncertainties.[32] Second, reflexive governance requires an experimental approach to policy interventions because "particular strategies, even if they appear to be the best solution from the perspective of current problem definitions, must ... be seen as hypotheses that are to be probed in practical interaction with the world."[33] Third, goals, though key as tools to create endurance, cannot be absolutely fixed in stone ex ante. Rather, the process of achieving deeper objectives is an iterative one, with goals themselves undergoing regular revision as knowledge and preferences update.

Institutions therefore need to build updating into their basic design to avoid the lag long problems cause. The most common tools for updating are mandatory review or sunset clauses. Review clauses require that an evaluation be carried out at a designated period, creating a potential trigger or hook for updating. Sunset clauses go a step further, creating an end date for a rule or institution unless it is proactively renewed. Institutional design could, as a principle for avoiding institutional lag, include review or sunset clauses and procedures to amend or reapprove rules in all first-order and second-order institutions, with more veto players in the former. Analogously, the goal-setting processes discussed in the previous section could also be included in regular review and updating processes, though goals should have longer time frames and higher thresholds for revision than the steps actors take toward implementing them.

Review and sunset provisions for laws or regulations have a long history. Plato advocates for them, and it was common for certain Roman edicts to take the form of *lex annua*, lasting only a year.[34] Medieval English kings forced parliament to limit the duration of its laws as a way to preserve royal prerogative and power. The nineteenth-century Norwegian constitution requires all tax provisions to expire each year unless actively renewed by the legislature.[35] More recently, sunset and review provisions have become more common for both legislative and executive rule-making as the regulatory state expanded in the postwar period.[36] The mid-century US political theorist Theodore Lowi advocated for a Jeffersonian "tenure of statutes" act that would put a five- or ten-year limit on regulatory agencies and their provisions to prevent an excessive accumulation of state power.[37] Still, empirical studies show that automatic review or sunset provisions are relatively uncommon.[38] Because they provide ways to limit the bounds of regulation, sunset clauses are sometimes linked to "small government" agendas. For example, in 2011, a new Conservative government in the United Kingdom adopted a "one in, two out" rule for new regulation, though the provision was later rescinded.[39]

Sunset clauses have several uses, not all germane to long problems. First, they are often applied to exceptional, temporary measures like states of emergency.[40] For example, during the COVID-19 pandemic, governments restricted people's rights to move and gather, but these limits were typically bounded in duration. Such provisions may be useful ways to adapt to temporary fluctuations away from and back to the status quo but not long-term successive shifts. Second, sunset clauses are used to facilitate legislative compromise or to dampen opposition. Soft opponents of a measure may balk at a permanent policy shift but may be willing to accept one for a limited duration. In systems like the United States where new rules face fiscal scrutiny, putting time limits on budget measures can generate a more favorable review, helping them pass even if their proponents expect them to become permanent. Indeed, in some cases, sunset clauses may become superficial. A policy's advocates may be able to keep renewing it with de facto automatic extensions, particularly for low-salience topics.

More relevant to fighting institutional lag, sunset clauses can help ensure that laws do not needlessly accumulate as conditions change. For example, eighteenth-century laws prohibiting witchcraft remained in force in a number of US states until the twentieth century. More problematically, state laws against homosexuality persist on the books even though federal jurisprudence has invalidated them. In the case of abortion, old state laws sprung back into force following the US Supreme Court's surprising reversal of federal protections for abortion in 2022.

For long problems, this inertia can be deeply problematic. Zoning rules provide a classic example. In many industrialized countries, rules were put in place in the postwar period to limit urban density. City planners, imagining a suburban, car-powered future, pushed development into single-family homes on large plots. Mixing residential and commercial spaces was prohibited. Restrictions were placed on multifamily housing and height. A generation later, major cities across the world face an acute, structural housing shortage. As prices exclude young, creative workers, cities' longer term economic prospects suffer. For these reasons, housing is an urgent and salient political issue across much of the world (New York even saw the emergence of the evocatively named Rent Is Too Damn High Party), but politicians find it very difficult to make substantive changes in zoning laws because the status quo bestows enormous benefits (in the form of high house values that have grown much faster than inflation) on those fortunate enough to be living the mid-century planners' dream. These groups tend to be older and wealthier, having become millionaires on houses that were available to the middle class in the 1960s–1980s and so wield significant political power.[41] What seemed like sensible urban planning in the past has now, in part through institutional lag, created a political economy that makes sensible urban planning extraordinarily difficult. The consequences of course extend far beyond housing. NIMBYism is now one of the most important political barriers to the changes in buildings and transportation that will be needed to address climate change. Had zoning rules come with review and sunset clauses, they would be significantly easier to modify as needs changed.

Sunset and review clauses can be seen as one way to generate what the political scientist Michael Manulak calls "temporal focal points."[42] One of the reasons institutional reform, particularly at the international level, is hard is because it requires a diverse array of actors to coordinate on how just what reform to agree but also when to pursue reform. Significant anniversaries can help overcome this coordination challenge by providing forcing moments that prompt updating.[43] For example, significant moments in international environmental governance have been marked by highly symbolic conferences that hark back to the 1972 Stockholm conference, such as the 1992 Rio Earth Summit, the 2002 Rio+10 summit, the 2012 Rio+20 summit, and so forth. Given scant space for reform on policymakers' agenda, temporal focal points like anniversaries or built-in review or sunset provisions provide informal mechanisms for generating the political conditions that can enable updating.

Sunset and review clauses help to build an adaptive, reflexive approach into governance systems that suffer from inertia. But it is useful to note that other approaches are also used. Although it does not tend to use formal sunset clauses, the Chinese governance system has an important adaptive capacity linked to the planning and target-setting framework described above. In essence, all big goals are reviewed every five years. Top officials set periodic, iterative goals that subordinates are then expected to meet. The exact means by which they are met can be flexible. If a certain law or provision is not leading to the desired result or the goal changes, the iterative planning process creates an automatic opportunity to adjust rules or goals. This can also enable a degree of experimentation and innovation at the local level, which can then be scaled up. As Zhao et al. describe environmental policymaking:

> The adjustment of targets and indicators in drafting each plan and the reform of policy instruments or implementation mechanisms facilitate adapting to changing domestic and international conditions, drawing lessons from previous plans and local experience, and soliciting suggestions and opinions from a wide range of actors including those in academia, think tanks, and internal government agencies.[44]

Helimann sums the distinctiveness of the Chinese system as "its unusual combination of extensive policy experimentation with long-term policy prioritization—i.e., foresighted tinkering."[45]

Of course, the authoritarian nature of the Chinese system interacts with this adaptive approach. Because the decisions and priorities of top leaders are authoritative and enforced via the review system described above, individual laws or regulations bend more easily to leaders' shifting priorities. When these priorities are well tuned to shifting circumstances, the malleability of Chinese rules and institutions can be an asset. However, it is important to note the trade-off of this system is the relative difficulty leaders then have to credibly bind themselves by locking in certain policies. For example, this can make it more difficult to enact long-term structural reforms. Chinese leaders have long desired to shift the economy away from an investment- and export-led model to an innovation- and consumption-led model. This requires, for example, clamping down on the tendency of local governments to support economic growth through excessive infrastructure investment financed by local land sales or financing vehicles, often relying on inflated real estate prices. Toward this end, the central government has frequently promulgated restrictions on local borrowing and sought to control white elephant investments and real estate speculation. However, when economic pressures bite—such as following the 2008 financial crisis or 2020 COVID-19 pandemic—the leadership reverts to prioritizing short-term economic growth and stability over long-term structural reform. In this way, the Chinese system's authoritarian nature risks making it too adaptive and thus too vulnerable to short-termism. From a climate perspective, this challenge is arguably the Chinese system's greatest weakness. Unless China's leaders can credibly commit to shifting the economy away from its current overreliance on inefficient investment in high-carbon infrastructure, they will struggle to achieve their long-term decarbonization goals.

At the international level, review is a common feature of multilateral rulemaking, and time-bound rules and institutions are relatively common. For example, under the SDGs, countries have agreed to review and update the subindicators under the seventeen goals regularly. In the UNFCCC process, many of the substantive rules have expiration dates. For example, the first commitment period of the 1997 Kyoto Protocol expired in 2012. Different bodies under the convention are often mandated into existence for five-year increments. For example, the high-level climate champions—tasked with mobilizing sub- and nonstate action—were created alongside the Paris Agreement in 2015 with a five-year mandate. This was then reviewed and renewed in 2019 for the period 2020–2025.

The ability to regularly review and update laws and policies (second-order institutions) is a sound governance principle at all levels. But what about reviewing and updating the institutions through which laws and policies are made (first-order institutions)? For example, should bureaucracies, legislative rules, and even constitutions and intergovernmental organizations have regular mechanisms for updating?

Many argue that they should be. Developing a more practical approach to Jefferson's idea of sunsetting constitutions after nineteen years, Jörg Tremmel proposes recurrent constitutional conventions that may amend constitutions at regular intervals.[46] At the international level, there are similar calls for reform. Most notably, many of the key international institutions today—the UN Security Council, the international financial institutions—embody the preferences and power structures in place after World War II. There has been limited reform over the years of the IMF and the World Bank, but the UN Security Council remains unchanged, save for the substitution of the Republic of China by the People's Republic of China.

Unlike laws or regulations, nearly all constitutions and international organizations have explicit updating mechanisms, which typically require special procedures and supermajorities, or in the latter case, the ratification of new treaties by member states. Such provisions reflect institutional designers' expectations that the "rules of the game" would need to update over time to ensure adaptability but that they should not be too easy to shift to ensure continuity. For this reason, changes to first-order institutions tend to have more veto players than changes to second-order institutions. Strikingly, however, no modern constitution or international organization has the equivalent of a sunset clause or even a review provision that can leaven their durability with adaptability. That is, there is no hook that requires political actors to consider modifying these institutions or to consciously renew them at regular intervals.

In a world of long problems, there is a greater need for malleability. Imagine, for example, the case of the UNFCCC. Although in theory countries could at any moment modify or agree new rules and procedures, in practice reform efforts are few and far between. Given the difficulty of getting the consensus needed to make a change, political leaders are loath to invest limited time and political capital. But the situation would be very different if countries knew that current arrangements would need active reapproval to continue. In this context, the potential to shape a new set of institutional rules and the risk of having unfavorable rules foisted on them would mobilize countries into action. In this way, review, sunset, and reapproval provisions could at least increase the potential for reform.

This benefit of course comes with a trade-off. By making institutions more malleable, we may weaken them. Imagine, for example, that the permanent members of the UN Security Council were to be updated every twenty-five years. Although the United States and the Soviet Union were able to agree a modus operandi in 1945, still united in defeating fascism, it is harder to imagine them coming together in the 1970 when Cold War tensions simmered and an active proxy war was unfolding in Vietnam. Similarly, it is difficult to imagine the United States agreeing a path forward with Russia and China today, just as it is hard to imagine the France or the United Kingdom (especially post-Brexit) giving their voice to the EU. If the creators of the UN had put an expiration date on the makeup of the Security Council, they would have set up a deeply challenging bargaining problem for their successors and perhaps raised the possibility that the institution they created would expire, with political actors too at odds to agree an extension. As Madison foresaw, the very existence of a review or sunset moment might "agitate the public mind more frequently and more violently than might be expedient." If the solution to institutional lag is more adaptability, the cost is some risk of institutional weakening or even breakdown.

An implication of this trade-off is that long problems will be better managed when political agreement around shifting not just second-order but also first-order institutions is more possible. As advocates of reflexive governance have written, "a key challenge is that deliberative and reflexive fora cannot be insulated from broader struggles for power and domination."[47] When distributional political conflicts are sharp or even existential, actors that benefit from present institutional arrangements will strongly resist changing them, and so institutional updating mechanisms will struggle to function. This inflexibility, in turn, raises the risk of dysfunction or breakdown as institutional lag bites. We see this clearly, for example, in the case of the UN Security Council. Because the stakes are high, the many efforts at reform have all been blocked. And because there has not been reform, the institution's effectiveness is diminished. A similar trend can be observed in countries before and after civil wars. Because the stakes are so high, neither side can make the kind of compromises in first-order institutions that could lead to a better outcome.

In contrast, rule changes have been more feasible in some of the international financial institutions. In the IMF, the World Bank, and similar bodies, voting shares between countries have been at least partially updated to reflect the increased economic heft of the emerging economies. There, the consequences of adjusting first-order institutions were less fraught, reform was not blocked, and the organizations are arguably more useful as a result. More acute political conflicts will undermine capacity for institutional updating and therefore for governing long problems effectively.

It must be recognized that this tendency bodes ill for efforts to govern climate change. The politics of climate change are growing more existential as the impacts of both climate change and efforts to mitigate it reshape broad swaths of the economy.[48] This means that we should expect it to be harder and harder to reform climate-focused institutions going forward, worsening institutional lag and perhaps driving actors toward less cooperative, more competitive strategies.

This challenge has a further implication for institutional design. Under conditions of greater political trust, actors will be more willing to accept processes for updating first-order institutions. If a country's citizens cannot trust each other to respect their basic rights, they will be loath to share power with them. The same will be true at the international level if countries worry that their peers will go against their fundamental interests. Social trust—or what at the multilateral level the international relations scholar John Ruggie termed an expectation of diffuse reciprocity[49]—can therefore meliorate the risk of changing institutions. Because long problems will require updates in first- and second-order institutions over time, they are likely to be managed more successfully in contexts with higher trust.[50]

Triggers and Reserves: Combining Durability and Adaptability

Long problems require institutional designers to take seriously both durability and adaptability as well as their combination. Two additional tools for doing so are policy triggers, instruments that automatically update once conditions change, and reserves, setting aside resources in the present to use later on. The former creates the commitment to act in the future; the latter creates the capacity to do so.

Trigger mechanisms abound in policymaking. Commonplace examples include indexing tax or spending provisions to inflation, revenues, or GDP or regularly adjusting legislative seats to changes in population.[51] Governments also set up triggers for more specific policy goals. Endangered species protections kick in once a population drops below a critical threshold.[52] In the monetary realm, countries often wish to ensure long-term stability for their currencies on global markets to promote certain economic objectives. When an exogenous shock drives currency values to high up or down, the central bank intervenes. According to the IMF, over 40 percent of countries use this kind of "managed float" regime for international exchange.[53]

Such tools can be thought of as "thermostatic institutions" because they involve both "(a) the automatic triggering of *swift changes* in one or more policy elements following some type of 'external perturbation' in order to maintain *durability* of other elements; (b) a high degree of *durability* to withstand (short-term) political and other pressures to eliminate the thermostatic institution itself."[54] In other words, such tools can prompt fast changes in service of long-term goals by eliminating future discretion.[55] In this sense, they can be seen as a specific form of the commitment devices described in chapter 4, carrying similar needs for and challenges around institutionalization to make them credible.

The COVID-19 pandemic provided another example of how triggers can be used. Following the first wave of infections and lockdowns, a wide number of jurisdictions began setting fixed criteria for what epidemiological conditions would trigger new closure and containment policies. Typically, these were expressed in metrics around levels or rates of infection and pressure on the health system like intensive care unit capacity. By committing themselves to a course of action ex ante, governments sought both to improve their reaction time while also reducing uncertainty about when citizens could expect new policies to be adopted, perhaps also reducing blame politicians received for unpopular measures. In practice, these triggers were not always followed, not least because the leaders did not formally constrain themselves.

Triggers are also a common way of managing reoccurring natural disasters like droughts.[56] Governments have created monitoring systems to assess the risk of drought. When these risks are detected, automatic triggers kick in, for example, to ration water use or provide support to farmers. By setting these triggers in advance, governments are able to provide longer term certainty around how droughts will be managed while also making themselves highly reactive.

Reserves are often used in conjunction with triggers and share many characteristics with them. Reserves are also perhaps the oldest and most common tool for long-term governance. Evidence of grain storage dates back eleven thousand years to neolithic communities in what is now Jordan.[57] Since humans began creating governments, food storage has been a major and universal concern. The *Book of Rites*, a core work of Confucian political theory, warns that governments will collapse unless they have at least three years of food on hand.[58] Today, economic interdependence means that supplies of basic goods are linked to global markets, creating both greater resilience (by diversifying supply) and new vulnerabilities and dependencies. Shocks like war can put access to vital resources at risk. For these reasons, societies continue to stockpile food, medicines, and fuel in particular through government arrangements, international pooling agreements, and market mechanisms. Reserves today rarely attract much attention, but they are vast, and their low salience is partly a tribute to their effectiveness at preventing the shortages that would capture political attention. Global stocks of wheat, corn, and rice are equivalent to one-quarter to one-third of global production.[59] Petroleum stocks of International Energy Agency member states are equivalent to a few years of their net imports.[60] These kinds of reservoirs will be helpful to address future impacts caused by climate change and will likely need to be expanded to promote resilience, particularly in places that lack the resources to build their own reserves. Indeed, some have proposed more multilateral reserves of critical resources that could provide for a more reliable and equitable safety net globally.[61]

Today, however, the asset governments hoard most is not commodities but capital. Government financial reserves take many forms, but from the perspective of long-term governance, one of the most interesting trends is the rise of sovereign wealth funds, a diverse category of government-owned investment vehicles. Early examples include the 1854 Permanent School Fund and the 1876 Permanent University Fund, created by the state of Texas to channel resources to address ongoing educational needs. In the twentieth century, as workers began to demand and receive more generous welfare benefits, governments set up pension funds to meet their future liabilities.[62] As populations have aged, the total assets of pension funds, both public and private, in OECD countries have grown larger than the combined GDPs of those countries.[63] More recently, export powerhouses like China and Korea have created funds to help them invest their foreign exchange surpluses more effectively. But today, perhaps the largest number of sovereign wealth funds, accounting for trillions in assets, hold the wealth that oil and gas producers have accumulated from selling the fossil fuels that create climate change. Norway, Russia, and the Gulf states, especially, have built enormous funds, as have entities like the state of Alaska. Analogously, Chile has created a large sovereign wealth fund capitalized principally by copper exports.

From the perspective of long-term governance generally and the climate challenge specifically, sovereign wealth funds can offer benefits on both sides of the durability-adaptability tension. First, related to durability, they serve as insurance against future risks. For example, many commodity-based sovereign wealth funds have been used primarily to manage the fluctuation in oil and gas prices, helping plug budget deficits when prices dip or, for example in the case of Russia or Iran, when sanctions bite. This relatively short-term price-smoothing helps ensure durability.

Over the longer term, capital reserves can to some degree substitute for the income from commodity extraction, helping ensure the durability of a country's fiscal health even as the original source of wealth, commodities, lose value due to exhaustion or asset revaluation. Norway created its enormous sovereign wealth fund in 1990, principally to aid with short-term price-smoothing. But as the fund has grown in the last decades, it has become an important source of revenue for the government, covering 20 percent of the budget despite conservative stewardship and spending rules. Government proceeds from ongoing oil and gas production still contribute more to Norway's public spending, but in the next decades, they will shrink considerably as fossil fuel reserves run out and climate targets bite.[64] In this way, the accumulation of capital reserves will continue to give Norway an enviable fiscal boost long into the future even as its original asset base dissipates.

Second, related to adaptability, sovereign wealth funds can invest in the long-term development of the country—for example, by funding infrastructure projects or industrial policy programs to reduce reliance on fossil fuel exports. In recent years, Saudi Arabia's Public Investment Fund (PIF), one of the world's largest, has shifted from a conservative, largely passive investment strategy to an activist approach. In 2016, the Saudi government launched its Vision 2030 campaign, a broad effort to expand the economy beyond fossil fuels. It correspondingly directed PIF to invest in domestic companies outside the oil sector and to pour funds into massive real estate and infrastructure projects at home, while also taking on higher risk investments abroad.[65] The fund's website prominently announces, "PIF is driving the growth of new sectors, companies and jobs, as a catalyst of Vision 2030."[66] The intention, challenging to realize, is to use reserves not to substitute for the future reduction in oil revenues (which Saudi Arabia can, and does, count on exploiting significantly longer than Norway) but to use them to actively build up other economic strengths.

Economic diversification is a challenging long problem that requires sufficient political support to sustain investment over many years and also to stand down or pay off the powerful incumbent industries and the ecosystem of stakeholders that depend on them (including, typically, governments themselves). To date, there are few successful examples, and reserves certainly do not guarantee success. But consider how much harder such a transition is for countries that lack a stockpile of capital to invest. Like Saudi Arabia and Norway, Nigeria's economy and government budget depend heavily on oil revenues, but its sovereign wealth funds amounted to only around US$2 billion in 2023. Both durability and adaptability will therefore be harder for Nigeria to achieve in coming decades as there is only a small reserve to substitute for fossil fuel revenues or to invest in building up alternatives do them.

As these examples highlight, in many contexts, triggers and reserves both create technocratic benefits (e.g., rationalizing policy, reducing transaction costs, managing risk) and serve political goals (like, for triggers, avoiding blame and limiting discretion or, for reserves, creating an alternative source of revenue to invest in political priorities or maintain political support).[67] They therefore tend to reflect contemporary power balances. For example, in the United States, social security payments (which benefit politically powerful older voters) are indexed to inflation but the minimum wage (which benefits politically weaker unskilled workers) is not; it must be updated through the discretion of Congress. And even though oil- and gas-producing states nearly all have the intention of reinvesting resource rents in diversification, in practice this has been hard because the political economy of rent-seeking creates powerful short-term political pressures that protect the status quo.

For long problems, triggers and reserves can helpfully combine durability with adaptability when changes can be anticipated and solutions to some degree preprogrammed. Shifts in currency markets, infection rates, or water levels can be expected, though there is uncertainty around when and to what degree they will arise. Solutions to future shocks can therefore be cued up in advance, ready to deploy when the need arises. For some kinds of institutional lag, such solutions will be very helpful indeed because they can overcome the uncertainty and inertia that impedes updating, while also reaffirming the longer term goals. We could envision these kinds of triggers applying to a much broader range of policy areas. For example, imagine if urban planning rules were automatically tied to house prices such that dips in affordability resulted automatically in policy changes that stimulated increasing supply and access.

For reserves to be effective instruments for tackling long problems, the same challenges around trusteeship, discussed in chapter 3, apply. Their governance must both mandate long-term objectives and insulate them from short-term pressures. Pension funds, for instance, are typically strictly regulated to ensure that they maintain their value long into the future. Similarly, the Norwegian sovereign wealth fund is required by law to limit annual dispersals to what it expects to earn in returns, about 3 percent per year. However, the Norwegian parliament could of course decide to change this rule at any moment, and use of the fund is actively debated in Norwegian politics. Similarly, the governance of the Saudi PIF depends strongly on the preferences of political leaders. As the energy transition becomes more disruptive and existential, it seems likely that pressure will increase on governments to use their reserves for short-term objectives that protect the status quo. Indeed, some research already shows, as we would expect, that sovereign wealth funds with greater insulation from political decision-makers tend to make longer term investments.[68] Giving these reserves more trustee-like governance protections can help resist these foreseeable trends.

Given these challenges, for triggers and reserves to be effective, policymakers need to be able to anticipate potential shifts in conditions and set up the right institutional mechanisms in advance. Could we imagine such a mechanism in the realm of climate? What if, for example, a rule had been in place in the late 1980s that linked the concentration of carbon in the atmosphere to a global carbon price? As emissions rose, an automatic correction tool would kick in, driving reductions. Alternatively, imagine that we had allocated a small portion of fossil fuel sales to a permanent, independent fund, perhaps administered by an international institution, that invested in low carbon technologies, providing a just transition for workers and communities that depend on fossil fuels, and responding to climate impacts. Although we cannot go back in time and create such a fund now, the implication is that we should ask ourselves what triggers and reserves could we create today to address the climate challenges of the future?[69]

Finally, policymakers facing long problems could make more use of triggers to link long-term goals to updating provisions. Consider, for example, a country pursuing a 2050 net-zero target through a series of interim milestones. A trigger could kick in if annual emission reductions failed to match the projected pathway, causing a review of the various policies aiming to cut GHGs. Instead of waiting for failure to drive sufficient political attention to a topic—with no guarantee that it would—trigger mechanisms could proactively ensure at least opportunities for updating, if not a guarantee that it would happen. Other triggers could be built in around interim targets themselves, creating opportunities to revise them as, for example, information on what is feasible or not feasible evolves.

Just like the goal-setting and updating provisions, triggers and reserves do not solve the tension between durability and adaptability for policymakers. Rather, they provide another way to institutionalize both the "good judgment" that policymaking processes strive for as they balance these two objectives, as well as the capacity to act in the future. To the extent long problems create more institutional lag, policymakers will need to rely on such institutions more heavily.

6
Studying Long Problems

This chapter explores how existing approaches in social science help us analyze long problems but also shows how taking problem length seriously challenges established theories, methods, and even the epistemology of political science and related disciplines. The chapter is more explicitly addressed to social scientists than the other parts of the book, so readers interested more in tackling long problems than studying them may wish to skim or skip ahead.

I argue, as the geologist Marcia Bjornerud puts it, "it is time for all the sciences to adopt a geologic respect for time and its capacity to transfigure, destroy, renew, amplify, erode, propagate, entwine, innovate, and exterminate."[1] Social scientists have made this case before.[2] The Nobel Prize–winning economist Douglas North offers some cross-disciplinary advice, "I will be blunt. Without a deep understanding of time, you will be lousy political scientists, because time is the dimension in which ideas and institutions and beliefs evolve."[3] Paul Pierson, who places that quote as an epigraph to his book *Politics in Time*, likens social science without time to cooking without regard to sequencing. You cannot just make sure you have the right variables in the right quantities, mix them up, and expect a delicious result; the order of steps matters. Alan Jacobs emphasizes that Harold Lasswell's famous summary of politics includes who gets what, *when*, and how.[4]

The chapter begins by showing how problem length helps us see issues differently. For example, it shows how mitigating climate change can be studied not as a free-riding problem but as a question of transition. The chapter then reviews some key theoretical and methodological tools social science offers to analyze long problems—a powerful kit. The heart of the chapter, though, argues where social science needs to go next: emphasizing rates of change as the outcomes of interest, accounting for dynamic problem structures, and analyzing future outcomes empirically. For each, I outline why this area is a fruitful realm for further exploration and, with reference to climate change, show how these ideas advance our understanding of long problems. Together, these three research strategies underscore the core argument of the chapter: long problems require social science to be more dynamic, future-oriented, and informed by the *longue durée*.

Taking the Period of Analysis Seriously

In chapter 1, I argued that problem length is a feature of all political issues. In this section, I show how it is also an analytic choice for the social scientist. Focusing on the length of problems invites the analyst to see them differently. Conversely, by not explicitly considering problems' duration, those studying them often make implicit assumptions that affect how we understand both political issues and their possible solutions. These temporal blinders can import the short-termism described in chapter 2 into our conceptual frameworks, theories, empirical research design, and explanations. The result is a common bias in social science toward shorter understandings of problems, which imposes temporal scope conditions on our work and make it less applicable, on average, to long problems.[5] To be clear, I am not arguing that all problems should be seen as long problems—length should be assessed empirically—but rather that social scientists should consider how problem length interacts with their frameworks and how it may render findings based on time-bound assumptions less generalizable over longer periods. Moreover, explicitly considering how political dynamics play out on different timescales can be a very productive way to examine problems from multiple angles and to compare different solutions, enriching social scientific theory.

Consider war, which social scientists study in both short and long ways. A canonical short approach is to model armed conflict as, essentially, inefficient bargaining.[6] States contest a certain outcome through a mixture of threats and inducements, and war breaks out if at least one side comes to believe it can secure its objective more effectively through force when the other party refuses to concede. The choice to go to war is of course also shaped by factors like uncertainty, the beliefs of elites, domestic political pressures, and bureaucratic politics, among others. Overall, however, this way of understanding war focuses on the moment of decision. Correspondingly, analysts highlight ways of reducing the threat of war that affect an actor's decision calculus at a point in time, such as deterrence, alliances, or collective security agreements to raise the cost of war for an aggressor, arms controls that remove certain capacities, or ways to credibly communicate a state's capacity and resolve to others, like weapons testing or forward deployments. These "solutions" take key aspects of problem structure, like the relevant actors, their preferences, and their capacities as given because at any particular point, they are essentially fixed. Other factors that can change on short timescales, like the information actors possess, then become the key variables of theoretical interest.

Other theories of war instead look at why states find themselves in conflict in the first place, focusing on factors like their internal beliefs and preferences (e.g., economic incentives to secure strategic resources), perceptions of other states (e.g., historical grievances or beliefs of injustice), or fears about their relative capacities in the international system (e.g., in the face of power transitions). These theories encompass a longer timescale, looking beyond the moment of conflict to understand the elements of the problem structure that led there. Under this expanded time frame, elements that are fixed in the short approaches, like preferences or power, are instead mutable and therefore the focus of these theories.

Theories of different lengths imply different types of solutions. Consider, for example, the observation that democratic countries essentially never go to war with each other. Social scientists continue to debate the mechanisms that explain this finding, and some have suggested short mechanisms, like the way democracies' underlying preferences and resolve are more transparent and legible to other states, decreasing the uncertainty around the likelihood and outcome of a conflict. Others, however, emphasize how shifts in preferences and norms associated with liberal, democratic states can create deep alignment of interests that can sharply reduce or even eliminate the threat of armed conflict, as we see, for example, in postwar Europe or between the United States and Canada. Such alignments, like the process of democratization, of course do not happen overnight. Indeed, Kant, who anticipated the democratic peace when liberal republics first began to emerge in the late eighteenth century, believed this outcome would take centuries to emerge as states democratized and learned to deal with each other. His version of the democratic peace hypothesis was not a static social scientific "law" but rather an evolutionary historical process.[7] So far, he is right.

As these examples show, armed conflict between states can be studied both as a long and a short problem. The different approaches emphasize different variables and solutions and ultimately complement each other. Explicit attention to the period of analysis helps students of these theories see their relative strengths and weaknesses and understand how they fit together.

Other topics also benefit from taking problem length seriously. For example, in the short term, addressing economic inequality is a matter of distributional politics: which economic interests can win power and direct policy to their benefit. But taking a longer term view instead focuses analytic attention on more structural factors such as the mechanisms of political representation— such as the changes in suffrage in nineteenth-century Britain that gave manufacturing interests more representation or the formation of labor parties in the early twentieth century.[8] Similarly, the problem of antibiotic-resistant diseases is, in the short term, a problem of controlling overuse and developing new medicines, essentially a collective action problem. In the longer run, however, new medicines will themselves become ineffective as microbes develop new resistance. Therefore, the "permanent" solution is to ensure the rate of development of new medicines stays ahead of the rate of microbe adaptation, which is a question of both controlling use over time and ensuring continuous research and development.

Climate change is of course a long problem. But theories of climate politics vary significantly in how they (often implicitly) account for the period of analysis. Strikingly, many leading social scientific approaches to climate mitigation, which I review below, have taken a short view.

Canonically, social scientists have approached climate change as a global collective action problem. This conceptualization emerges from theories of natural resources and collective goods, and begins from the observation that the capacity of the climate to absorb GHGs is, like a shared pasture, both rival (my use of it diminishes your ability to use it) and nonexcludable (there is no inherent limit on who can use it). Such "common pool resources" are prone to overexploitation because each actor has no incentive to limit its own use. After all, if others refuse to hold back, an actor will scarcely benefit from being the only one acting sustainably; the resource will still deteriorate. Alternatively, if everyone else limits their exploitation to a sustainable level, any self-interested actor will have an incentive to "free ride," to take advantage of the restraint of others to enjoy more of the resource for itself without having to worry about the consequences of overuse. Because this logic applies to everyone, creating a large-scale prisoner's dilemma, overexploitation follows.[9]

Because acting is costly, actors will need some sort of inducement or incentive to do so. For example, if an actor knew that by acting, it could be assured all others would act (say, because they all agreed to a global treaty with hard monitoring and enforcement provisions), then the cost of acting would be worth it because it would unlock reciprocal action from others, producing the collective benefit.

As applied to climate change, assumptions about problem length play a key, though typically implicit, role in this argument. This classic model of collective action is static, offering a snapshot of incentives at a particular moment in time. But when applied to a long problem like climate mitigation, we need to assess how much value actors place on the *future* benefits of collective action. After all, the benefits of reducing emissions materialize only later on, so the "shadow of the future" has to be worth it. Actors with longer time horizons will therefore have more incentive to act now, all else equal.[10] Unfortunately, as the discussion around short-termism showed, few actors are likely to have such long time horizons. In this way, short-termism helps justify the canonical interpretation that no actor has an individual incentive to act absent some collective agreement because, on average, the costs of action today outweigh the value derived from the future collective benefit.

Put another way, in the classical model, collective action on climate mitigation depends not just on overcoming the free-riding problem but also assumes that actors must sufficiently value the future if they are to act in the present. In this way, the "tragedy of the commons" is equally a "tragedy of the horizon."

Some scholars question the assumption that free-riding concerns lie at the heart climate politics, though in some ways they share a short view of the climate mitigation problem. For example, Ostrom and others have posited a "polycentric" model of climate change in which mitigation is best interpreted not as a single global problem but rather as a multitude of interlocked but distinct problems that states, cities, businesses, and other actors face: building public transportation, electrifying the vehicle fleet, taxing fossil fuels, and so forth.[11] At these varied scales, the polycentric view argues, costs and benefits may look quite different, and collective action may be easier to grow and sustain since it can latch on to "thick" networks and institutions that exist at more granular scales, as opposed to "thin" global governance. The polycentric viewpoint therefore cautions against theorizing climate change through a single unified problem structure and instead urges analysts to break the larger, abstract problem into smaller, concrete subproblems. Importantly, though not a focus of this theory, some of these disaggregated problem structures may be shorter than mitigation writ large. For example, reducing air pollution improves the health of people alive today.

In a similar vein, the political scientists Michaël Aklin and Matto Mildenberger critique the collective action model by arguing that there is little empirical evidence to support the idea that concerns over free riding significantly affect countries' actions on climate change, or lack thereof.[12] Instead, they emphasize the distributional struggles within countries over climate policy. This distributional focus emphasizes who wins and who loses from climate policy and how these competing interests contest for power and influence. This approach focuses very much on the present—who has power and what are their preferences? It deemphasizes the role that future collective benefits might play in unlocking action by arguing that, consistent with short-termism, the most important variables are the balance of power and interests in the present. By this logic, we can surmise that contra the classic model, even if a collective commitment to unlock future benefits could be agreed, it would matter little if anti-mitigation interests are in power.

Each of these three approaches to understanding the climate mitigation problem—the classic collective action model, polycentricism, and the distributional approach—takes a short view. They do not theorize problem length explicitly, though the classic model assumes some degree of long-term value must be present for collective action to be worth it. For certain research questions, like explaining historical episodes of climate policy, this may not be a problem. But to overcome temporal scope conditions, particularly important for long problems like climate change, the full theoretical toolkit needs to theorize problem length explicitly.

Consider, again, the canonical collective action approach. A long view invites the social scientist to question some of the strong, static assumptions in the classic model. For example, does it make sense to model the costs and benefits of action as constant? After all, we can see in many areas of climate mitigation that costs and benefits change over time, and some costs and benefits are in part functions of the total level of action. For example, as discussed in chapter 3, the willingness of early leaders like California, Denmark, Germany, and Japan to subsidize the early research and especially deployment of solar and wind technology fundamentally reshaped the costs and benefits of deploying renewable energy.[13] These kinds of increasing returns mean that the incentive of an actor to act is not just a function of costs and benefits in the present but of previous action as well.

Combined with the willingness of some actors to take some actions unilaterally (following the logic of the polycentric and distributional approaches described above), such increasing returns create the possibility of what I have called catalytic cooperation.[14] In this approach, the core challenge is not overcoming free riding but spurring a critical mass of early action such that the incentives actors face for further action become more favorable. In some cases, if increasing returns to action are strong enough, tipping points may be reached at which acting becomes strictly preferable to not acting. In such scenarios, collective action flips from being difficult to achieve to hard to avoid. To the extent such dynamics apply, they show how the pessimistic conclusions of the classical model of collective action depend in part on certain assumptions about the fixed nature of preferences and so may suffer from temporal scope conditions.

The catalytic cooperation model, whose policy implications chapter 3 discussed, is one example of the large literature that sees the climate problem through the lens of transition. These approaches come in many flavors but share a common focus on explaining how change (or lack of it) emerges over time. To cite some examples, the transition scholar Frank Geels takes a structural view that considers how innovations in technical, economic, or social systems can emerge and scale up over time, both endogenously, but also due to exogenous windows of opportunity.[15] Political scientists Joshua Busby and Johannes Urpelainen focus on the leader-follower relationship, seeking to understand, once leadership has begun to emerge, the conditions under which others follow or not.[16] Environmental governance scholars Kelly Levin, Benjamin Cashore, Steven Bernstein, and Graeme Auld, building on their analysis of climate change as a "super wicked" problem, focus more on the dynamics that make scaling up possible, arguing that interventions should be "sticky" (hard to reverse), leading to greater entrenchment, and covering expanding populations over time.[17]

Steve Bernstein and Matthew Hoffmann, leading thinkers on global environmental governance, take a similar view but focus more on overcoming the "lock-in" that make it difficult for innovations of interventions to get off the ground.[18] In line with the polycentric approach, they advocate breaking the broader climate problem into subproblems but argue that these subproblems can be thought of as a fractal. Each shares a common pattern of carbon lock-in, and cumulatively, interlinkages between the subproblems replicate carbon lock-in across the system as a whole. From this analysis, it follows that decarbonization is the process of disrupting carbon lock-in in more and more subproblems, gradually tipping the system as a whole out of self-reinforcing high-carbon systems.

Finally, the distributional view of climate politics can also be complemented by a longer perspective, as the political scientists Jeff Colgan, Jessica Green, and I have put forward in our theory of asset revaluation and existential politics.[19] The argument begins from the premise that both climate change and decarbonization unfold over time. As the former proceeds, climate-vulnerable assets (e.g., farms, coastal land, etc.) will lose value, and eventually, those who hold them will lose political power. Conversely, to the extent decarbonization proceeds, the same collapse of asset value and ultimately political power will befall climate-forcing assets (e.g., coal plants, steel factories, etc.) and their owners. As both trends advance, certain assets will be threatened with extinction, creating an extreme case of distributional politics we term existential politics in which the nature of political contestation changes. As an increasing range of actors hold progressively more intense preferences over climate policy, politics shifts from a question of who gets what to a question of who gets to survive.

As these examples show, approaches to theorizing the politics of climate mitigation vary in how they incorporate problem length, even when they share other attributes. Both the classic free-riding model and the catalytic cooperation model emphasize collective action, but the latter relaxes the static assumptions of the former. Both polycentrism and the fractal lock-in approach break the larger problem into subproblems, but the latter considers how the self-reinforcing dynamics of lock-in can be broken down over time, consistent with other approaches in the transition family. Both the distributional politics approach and the asset revaluation approach center on contestation for power, but the latter theorizes what implications winning or losing those contests at one point in time may have for who wins or loses at later points in time.

These different theoretical approaches suggest different kinds of solutions. Because most social science theories identified free riding as the main problem, two categories of solution have tended to be prominent. First, as noted above, a collective agreement through which all commit to limit their use of the resource to a sustainable level conditional on everyone else making the same commitment. If such an agreement can be forged and assuming monitoring and enforcement are sufficient to make it credible, each actor would be strictly better off in the long run because the resource can be maintained into the future. This logic animated the design of the 1997 Kyoto Protocol, whose drafters were highly influenced by the mental model of the 1989 Montreal Protocol targeting ozone-depleting substances (in many cases, it was the same individuals involved).[20] Second, free riding might be overcome by creating some kind of climate club around an excludable benefit. For example, many scholars have proposed tying deep emissions reductions to trade measures that would raise the cost of imports from high-carbon jurisdictions. If a critical mass of countries adopted such a policy, conditioning market access on climate action, free riders could be deterred through the prospect of strict enforcement.

On the other hand, if the problem is not so much free riding but rather how to generate a critical mass of action in the present to shift incentives in the future, then different solutions are needed. As early as 1988, Steve Rayner (an Oxford don who described himself as an "undisciplined social scientist"), drawing on early international relations regime theory, proposed a polycentric system including both state and nonstate actors experimenting with a range of voluntary solutions and sharing information about success and failure through loose networks.[21] Following the catalytic cooperation model, the "catalytic institutions" described in chapter 3 seek to build a critical mass of action over time by stimulating first movers through flexibility, iterating targets over time, and increasing the effect of prior action on subsequent action through mechanisms like normative benchmarking, information-sharing, and capacity-building. Aligned with these ideas but more skeptical of the Paris architecture to implement them, Charles Sabel and David Victor's experimentalist governance model, also discussed in chapter 3, presents a process through which innovations can be seeded, encouraged, and scaled to the point they replace the status quo.[22]

Finally, while the distributional approach and asset revaluation model share a common focus on contestation for power, their different treatment of problem length implies some different solutions. For the distributional approach, winning power is key, perhaps through coalition-building. The asset revaluation approach, in turn, suggests that solutions should be focused on overcoming obstructionism. Winning power may provide one way to do this. If one side can win power and devalue the assets of the other, it will remove the root of their political power. But it may also provoke a backlash as climate-forcing asset owners, their backs to a wall, resort to more extreme forms of contestation (up to and including the use of force). An alternative pathway could be to "neutralize" opponents of climate policy through compensation, buying out and retiring their assets. "Paying the polluter" may seem like a normatively disagreeable policy choice, but an asset revaluation framework shows how it could overcome obstructionism. Consider, for example, the way in which slavery was ultimately abolished in the British colonies in the 1830s. In a series of laws, parliament effectively paid slave owners, who had fiercely resisted change, the modern equivalent of billions of dollars in compensation for the loss of their "property." Could similar forms of compensation overcome opposition from oil companies or coal miners?

As the different understandings of climate mitigation sketched above show, explicitly considering problem length both demonstrates how different approaches relate to each other and emphasizes different kinds of solutions. The above arguments focus on mitigation, but as noted above, that lens also truncates the climate problem in important ways. If we see climate change as an adaptation problem, then we need to consider not just the timescale on which decarbonization may take place but also the timescales on which different kinds of interventions to address the impacts of climate change may unfold. Some, like long-term infrastructure investment, may have significant gaps between cause and effect. In other words, by focusing on mitigation, existing social science approaches have already (perhaps implicitly) put temporal scope conditions around their understandings of the climate problem.

The argument is not that short approaches are wrong—indeed, they often provide clear and parsimonious explanations for immediate questions. But short approaches are less generalizable across time, which can be a limitation for the study of long problems. We should ask not just how and under what conditions a certain explanation obtains but also *when* and *for how long* it obtains. Fortunately, social science offers a wide array of theories, concepts, and analytic tools to grapple with these challenges. The next section explores them.

Existing Tools

Long problems are not new to social science. An array of methods and theories help us understand problems where causes and effects span long periods and some, like historical institutionalism, take time as a central focus. Before the rest of the chapter proposes where social science can go next, this section highlights some of the most essential tools in our current kit. Readers already broadly familiar with common social science approaches may choose to skip it.

Starting from approaches that theorize actors' behavior and interactions, a key question raised by long problems is what preferences actors have at different points in time. In short understandings of problems, preferences tend to be fixed; it is information or the actions of others that change. But longer approaches require us to ask how preferences might vary. For example, chapter 2 noted the issue of time inconsistency as part of short-termism; individuals, corporations, or nations may say they will do one thing today but then do another tomorrow. A politician may promise to increase school spending before an election but then face difficult fiscal trade-offs once in power. Time inconsistency therefore undercuts actors' ability to make credible commitments, which in turn can make governing long problems difficult. For example, why would a politician invest in a pension fund now if she expects some future government to raid it for their political advantage? Institutional solutions like goal-setting, discussed in chapter 5, are therefore needed to address the challenges of shifting preferences from one period to the next. Whereas preferences can be inconsistent within a short span of time, we can assume that, on average, inconsistency increases as problem length increases.

Also related to preferences, social science seeks to model how much, or little, actors value the future compared to the present. In economics and other fields, scholars rely on discount rates for this purpose. In narrow financial applications, the discount rate reflects in part the idea that a dollar today is worth more to you than a dollar in the future. More broadly, discount rates can also reflect so-called pure time preferences (you may really just want that dollar today, not tomorrow) and greater uncertainty about the future.

Although useful for some applications, discount rates are difficult to operationalize or measure in practice. They model time by flattening it into a single parameter, a very blunt assumption that can lead researchers astray. Psychological evidence suggests that most individuals' irrationally discount the future in a hyperbolic pattern, meaning that individuals' discount rates are probably quite low into the immediate future but then ramp up precipitously and remain relatively flat long into the medium- and long term. Summarizing such preferences in a single average discount rate can be a useful shortcut for modelers, but it can be highly misleading when, for example, a modeler decides by assumption that longer term outcomes are less worthwhile than shorter term ones.[23] Moreover, given that we expect preferences may change over the course of a long problem, a constant discount rate is unlikely to be theoretically satisfactory.

But even if we cannot usually measure discount rates and risk misusing them, it is valuable to understand variation in the time horizons of different actors, not least so that we can apply the kind of horizon-shifting strategies discussed in chapter 4. Why do some value the future more or less or value it sometimes but not others? Social science emphasizes how institutional and behavioral factors shape such preferences. As noted above, short-termism is often a product of regular evaluation cycles. Actors that are more insulated from such pressures, such as those that do not face immediate electoral punishment, tend to have longer time horizons than those who do not. Thus, politicians are more likely to make sensible investments in pensions when they do not face electoral risks, and governments take more energy efficiency measures when they are protected from consumer backlashes.[24] Similarly, institutional investors mandated with, say, maintaining university endowments or pension funds in perpetuity will have very different time horizons than other market actors. Political parties characterized by "overlapping generations"—that is, those that have up and coming leaders who hope to rule long into the future—will invest more in long-term goods than parties that are only transitory.[25] Noninstitutional conditions also affect time horizons. Studies show that poor people are less able to plan for the future than richer ones since they must spend relatively more cognitive energy maintaining their precarious economic position each day.[26]

Social science also has tools to address questions of timing, sequence, and contingency. Many social phenomena—notably, political institutions—are characterized by path dependence, which Pierson and others define specifically to refer to processes of increasing returns.[27] In such processes, a certain choice becomes progressively self-reinforcing over time through positive feedback mechanisms. For example, the advent of mass literacy in the nineteenth century allowed countries to train citizens in national languages and identities which then substantially reinforced nation-states.[28] Processes may also exhibit decay instead of reinforcement, but in general, we tend to observe increasing returns rather than decreasing ones, as the former persist while the latter fade away.[29] Path dependence therefore helps us understand the conditions under which the elements of long problems sustain themselves over time.

Critical junctures provide another related way to think about both how earlier things affect later things and also how the choices available to political actors may differ over time.[30] Critical junctures are "relatively short periods of time during which there is a substantially heightened probability that agents' choices will affect the outcome of interest."[31] During a critical juncture, structural constraints may be broken, shifting, or malleable such that actors can make choices that will open up or close off certain future possibilities. For example, in the aftermath of major wars, victors have typically established new institutional arrangements that substantially reorient the international system.[32] Such outcomes are possible typically because the constraints around political choices have shifted. For example, a crisis may have weakened existing structures that previously blocked off certain choices, like the collapse of a great power such as the Soviet Union in 1989. Alternatively, a "temporal focal point" (like a major anniversary) may emerge that allows actors to overcome coordination costs,[33] or a critical juncture may arise simply because of long-running shifts in structural background conditions. However a critical juncture emerges, once choices are made, they are quite difficult to reverse because of path dependence or other mechanisms. When studying long problems, it is important to understand whether the period of analysis includes any critical junctures or not. Moreover, long problems like climate change, which can cause crises, may generate critical junctures endogenously.[34] Left unaddressed, long problems can metastasize into transformative crises following a dynamic of punctuated equilibrium.

Path dependence and critical junctures highlight the role of what Pierson terms slow-moving causal processes.[35] For example, cumulative problems like climate change require time to register an effect. Sometimes such processes exhibit threshold dynamics in which little change is registered for long spans until some critical threshold is reached. This pattern has been identified, for example, in normative changes, when ideas like abolition, women's suffrage, or gay rights go from fringe to suddenly mainstream.[36]

Finally, critical theorists have drawn attention to the study of time and particularly its relationship to power. Liam Stockdale, for example, argues that focusing on future risks like a potential terrorist attack empowers actors who claim to predict and prevent them, such as intelligence agencies.[37] Anticipation is not neutral but instead shifts power to those who determine what the risk is and what steps might be needed to mitigate it, including, in Stockdale's examples, preventative war. In this way, future risks provide a rationale for ongoing control measures or other exercises of power in the present. In a similar vein, Jasanoff and Kim show how scientific and technological visions of the future carry with them stark implications for public priorities and purposes, deeply political choices that can reorder power in the present.[38] This critical lens may be particularly useful for understanding the implications of defining problems in long or short ways by focusing attention on how different constructions of a problem create different power dynamics.

As this brief survey demonstrates, we already have many ways of theorizing how political preferences, institutions, behavior, and power can vary over extended periods. Of course, operationalizing these concepts requires methodologies to study their mechanisms empirically. Again, a significant set of existing tools can be used here.

Historical institutionalist approaches often use qualitative process tracing to explore dynamics like path dependence and critical junctures.[39] Using these methods, scholars identify specific causal mechanisms that, for example, generate increasing returns, determine their observable implications, and assemble qualitative evidence to determine if the posited dynamics obtain or not. Related, analytic narrative approaches use formal theory to outline specific causal processes within a given case, which can then be used as a framework for qualitative exploration.[40]

Quantitative approaches also offer ways to study time trends. Many statistical analyses seek to control for the effect of time with, for example, splitting panels at different break points (in international relations, it is common to include a dummy variable for before and after the end of the Cold War) or through various controls that seek to account for autocorrelation: the idea that the value of a random variable at any given point in time will be strongly affected by the previous value of that variable. These methods are commonly used to try to get rid of the causal effects of time to pinpoint more accurately some hypothesized static effect. But such approaches can equally be used to explore the role that time plays in a causal process—for example, in the study of diffusion—where a key mechanism of interest may be how the number of other actors adopting a certain policy affects the probability of another actor to adopt it.[41]

Finally, various forms of modeling can help us explore long problems. Models offer a structured way to create scenarios or falsifiable hypotheses from a starting set of assumptions and parameters via internally consistent logic. In this way, modeling can help us identify causal processes to explain both past events[42] and to develop systematic expectations about future outcomes, particularly helpful for considering long problems (I return to this point below). Models can be built from empirical observations or theoretically motivated assumptions, or a mix of the two.

In formal theory, game theoretic models are particularly useful for exploring ideas of sequencing and conjuncture. In situations with multiple possible equilibriums, the order in which things happen is often decisive for determining outcomes. In a similar vein, agent-based modeling can add a stochastic element to this kind of analysis. In this approach, the researcher defines the preferences and possible actions and strategies of a number of actors and then uses a computer simulation to see how they behave over a certain time period and with what overall outcomes.[43] It goes beyond formal theory in that outcomes are determined not only by assumptions but also by many variable interactions.[44] For example, we can program a simulation in which we give actors preferences (e.g., maximizing short-term economic growth), influence (perhaps weighting them by the size of their economies), and choices (cooperate or not cooperate) and then allow them to interact in multiple rounds to see if they generate more or less cooperative behavior.[45] While such models cannot of course predict the future—they are only as "right" so far as their built-in assumptions obtain—they can help us understand how individual parameters matter. For example, a simulation in which actors place little value on the future can be compared with one in which they value it highly.

Outside the bounds of social science (as conventionally understood), researchers use other kinds of tools to examine long problems empirically. In the realm of climate change, integrated assessment models (IAMs) provide the primary way policymakers and publics understand the problem.[46] These are the tools scientists rely on to estimate the changes in global temperature that will result from different courses of action. When the IPCC or other bodies state that the world is not on track to achieve the goals of the Paris Agreement, it is ultimately IAMs that form the basis of these estimates. IAMs are large, complex computational models that seek to capture both natural and human systems, parameterized with observed data. Although IAMs are not typically seen as social scientific tools, they work by making certain assumptions about, for example, the level of economic growth and the degree of technological change over a number of decades, or the policy choices different actors make. In this way, they reflect assumptions that depend on economic, social, and political outcomes and so are, de facto, making strong claims about the dynamics social scientists study. Correspondingly, there are various attempts to more explicitly model some of these dynamics in IAMs, but such efforts have not yet proceeded particularly far. Even obvious insights from social science, like the notion that policy implementation depends in part on state capacity, are not typically reflected in IAMs.[47]

As computational models, IAMs are transparent and rigorous in the sense that their assumptions and parameters can be clearly stated and compared and their internal validity is precise. However, for many users—not least policymakers—their complex, quantitative nature makes IAMs difficult to assess,[48] and their macro scale gives them poor resolution on many of the core social science questions and policy challenges long problems raise.

Modeling is of course not the only way to generate scenarios. As discussed in chapter 3, in both the business and policy realms, a large practice (and associated consulting industry) has arisen around foresight, scenario planning and scenario analysis, essentially a family of approaches (quantitative, qualitative, and mixed) to think about plausible future outcomes.[49] Unlike models, scenario planning does not necessarily derive outcomes from starting assumptions that flow through internally consistent logics. Rather, it seeks to identify more or less plausible future pathways and to consider their implications. The aim of scenarios is not to predict what will happen (sometimes called forecasting) but rather to explore what may happen in a structured way to inform decision-making in the present.

Even further from the core of social science, there are other ways to think about long-duration problems—for example, simulation, gaming, and storytelling. These creative, nonpositivist approaches do not seek to explain social phenomena but rather to generate visions of potential futures. Whereas scenario-based approaches use heuristics and structures to assess possible futures, these creative, humanistic, techniques instead rely on imagination, narrative, psychology, and interpersonal relationships, though many include sophisticated treatments of social and political dynamics. Militaries have long used war-gaming to facilitate their planning efforts. Climate change, and particularly its future impacts, has become of a major subject of art, film, literature, and music—in turn prompting an extensive scholarship in the humanities and in sociology examining the meaning of this cultural interest.[50] Such work helps us explore long problems not only through analytic tools but via emotion, affect, and imagination as well.[51] One intriguing possibility is that these humanistic approaches may allow us to develop some degree of empathy with future people and so lead us to make more long-term choices today.

In sum, both social science and ways of thinking that fall beyond its boundaries have many tools with which to address long problems. But it needs to go further. Below, I outline three ways forward and show how they help us understand climate change in particular: focusing on rates of change, looking at dynamic problem structure chains, and studying future outcomes empirically.

Rates of Change

The longer a problem, the more social scientists should be interested in rates of change as opposed to levels. Imagine a race in which each runner has a slightly different starting line. If the contest is a hundred-meter sprint, then an advantageous starting position (level) can be decisive. But if the race is a 10k, a small difference in starting position will be insignificant and the runner's speed (rate of change) will be what determines the outcome.

Long problems are marathons, and so rates of change and their determinants are likely to be key variables of interest. Recall the example of short and long approaches to war from the start of the chapter. The determinants of relative power depend on the length of the period of analysis.[52] If we consider a single battle, we should study the number and quality of military assets both sides possess. But if we want to know the outcome of an extended war, it is the speed at which a country can produce armaments and train soldiers that matters. For example, at the start of World War II, the navies of Japan and the United States were roughly equivalent, but Japan believed the United States, with its larger economy and greater access to resources, would outproduce it over time. Hence, the Japanese military leadership became convinced that a preemptive strike on the US Navy was necessary, resulting in the 1941 Pearl Harbor surprise attack.

If we consider an even longer problem, such as an extended period of geopolitical rivalry, it is not just the speed at which a country can build weapons and train soldiers that matters but the overall rate of economic growth or technological development that underlies production capacity. For much of the early Cold War, the Soviet and Western militaries were similarly matched or the balance slightly favored the Soviets. Soviet factories could produce nuclear weapons on a scale to match or outpace the West. But over the decades that followed, Western militaries, buoyed by growing, technologically sophisticated economies, developed decisive advantages and were able to innovate more quickly. Over the long haul, it was the rate of change in the rate of change in military power that mattered most.

Extending this idea, in Liu Cixin's famous science fiction trilogy *The Three Body Problem* an advanced alien species hundreds of light years away attacks Earth's scientists.[53] Even though human technology is vastly inferior, the aliens are terrified by our rapid rate of change. Humans went from the Stone Age to the Information Age in a mere ten thousand years, while on the aliens' world, a similar transformation took eons. The aliens fear that in the hundreds of years it would take them to travel to Earth, our technology would develop to outmatch theirs, and so their "Pearl Harbor" strategy targets not our military but our ability to learn: a long-distance attack on Earth's scientists.

The importance of rates of change in long problems has three implications for the social scientist. First, we need to understand how the rates of change of different elements of a long problem compare to each other. Indeed, we can often define long problems in terms of this relationship. Over time, faster factors may come to dominate those that operate more slowly, like weeds overrunning a garden. For example, it has been hypothesized that warm-blooded animals like humans have evolved to have body temperatures higher than the external ambient temperature to protect ourselves against microorganisms. An organism used to living outside will struggle to survive in our warmer bodies. However, scientists now worry that with climate change, fungi in particular may be evolving to survive in warmer and warmer temperatures, allowing them to exist quite happily inside our bodies with dangerous consequences.[54] Over time, mammals may very well evolve to run hotter body temperatures to compensate, but we reproduce and replicate far more slowly. In this way, the faster rate of evolution in microbes creates a real risk to humans. Similarly, the problem of resistance to antibiotics is a problem of relative rates. At any level of antibiotic usage, harmful pathogens will begin evolving in ways to overcome our defenses. The question is whether the rate of resistance development (which is driven by our use of antibiotics) exceeds the rate at which we develop and deploy new medicines. The problem can be understood as managing the relationship between two rates of change.

Second, if rates are the outcomes of interest, then the factors that affect them—accelerators or decelerators—become key variables. In mathematical terms, we should pay attention to not just the value of a variable and its rate of change (first derivative) but also the factors that shape the rate of change (second derivative and potentially beyond), as in the military examples above.

Taking this approach to long problems allows social scientists to avoid, for example, the mistake that the early demographer and economist Robert Thomas Malthus made in his famous argument about population growth and economic development. In his 1798 *An Essay on the Principle of Population*, Malthus argued that economic growth leads to population growth, which ultimately keeps living standards low. Overall, societies did not get richer; they just got bigger. This was primarily an argument about levels (a given quantity of economic growth would lead to a commensurate quantity of population growth) that made strong implicit assumptions about rates of change being fixed. In Malthus's world (which he believed reflected divine design), the speed of economic growth can never decouple from the speed of population growth. In other words, the Malthusian trap can be seen as the assumption that the first derivative of two variables, economic growth and population growth, are linked such that one cannot outpace the other. But Malthus turned out to be wrong because his theory omitted (among other things) a critical driver of the rate of change in economic growth: technology. Fertilizers and mechanization allowed the same amount of resources to produce a lot more food, radically expanding the rate of food production above that of new births, and creating not just growth but, at least for some, abundance and wealth. And with further industrialization, it turned out that rising standards of living can actually lead to reductions in the birth rate, turning Malthus's trap on its head.

Students of long problems need to be attentive to these kinds of dynamics to avoid Malthus's mistake. We should consider not just the levels of key variables of interest and their rate of change but also how the rate of change may itself change. Of course, in extremis, this approach could lead to analytically unhelpful, potentially infinite regress, as the social scientist seeks to determine the causes of the causes of the causes. To paraphrase an expression, it is derivatives all the way down. The point is not that analysts must always dig as deeply into the causal chain as possible. Rather, we should recognize that the value of such digging increases with problem length. The longer the problem, the greater the need to consider the deeper determinants of rates of change. For a long problem like climate change, it is difficult to see a satisfactory way to study it without considering rates of change in the economy and technology, and so understanding what affects these factors is key to understanding what degree of climate change we experience. For example, the International Energy Association has consistently underestimated the global deployment of solar and wind technology because it used an overly linear model to predict future changes in the price of renewable technologies, which does not adequately account for the way prices fall faster as deployment speeds up and reaches a larger scale.[55]

Third, a focus on rates of change means we should think about "solutions" to long problems not as a fixed end point but rather as achieving a positive direction of travel and (relative) rate of change. That is, policy interventions should focus on speeding up or slowing down as much as on stopping or starting. At any given point in time, it is difficult to fully resolve a problem whose effects manifest only long into the future. But policymakers can at least make interventions that put us on a path toward addressing a problem. Continuing with the example of antimicrobial resistance, we will never end this problem because microbes will always continue to evolve. However, by reducing our use of antibiotics, we can slow it down, and by investing more in research into new medicines, we can accelerate the pace at which they become available. Solving antimicrobial resistance can therefore be thought of as keeping the rate at which pathogens evolve permanently below the rate at which we develop new medicines.

Thinking about climate change as a problem of rates helps us see it more clearly. We commonly define the climate problem in terms of levels—for example, limiting warming to well below 2°C or 1.5°C, as the Paris Agreement states, or bringing carbon dioxide levels in the atmosphere back to something like 350 parts per million. But such statements are incomplete without also specifying a time dimension. The climate has always, and will always, change on a geological timescale. The problem is that now anthropogenic emissions have radically accelerated the natural rate of change (which is itself not constant). Zooming in on just recent atmospheric history, it took about a century from the start of the Industrial Revolution (1880–1980) for the climate to warm by 0.5°C. But it then took but less than half as long (1980–2015 or so) to warm by the same amount again. We are on track to hit the next half degree even more quickly. This rate is about ten times faster than the warming the earth has experienced between the last Ice Age, twenty thousand years ago, and the start of the industrial era. Over millennia, a few degrees of temperature change is a major shift. Compressed into a single century, it is a planetary catastrophe.

If we look at climate change as a problem of rates, the three points outlined above—comparing different problem elements' rates of change, theorizing factors that affect rates of change, and seeing solutions as speeding up or slowing down—give us new insights.

First, consider the relative rates of change between the mitigation and adaptation aspects of the climate challenge or, roughly, between prevention and treatment. Anthropogenic climate change is a problem because its speed vastly exceeds the rate at which natural and human systems can adapt to it. If human and ecological systems were able to somehow quickly shift people, agriculture, and flora and fauna around the planet, to instantly develop the ability to survive extreme heat and weather, to live equally well on land or water, or otherwise adapt at speed and scale, then rapid climate change would be no more threatening than a change in the weather. In reality, though, human systems and the natural systems they depend on are slow to change. We have built capital-intensive settlements, mostly in coastal regions. We are tied to certain places by culture, identity, property rights, and sovereign borders. We cannot quickly change our biology to endure harsher weather conditions. Even lifesaving technologies like air-conditioning or seawalls, which could be deployed relatively quickly at scale, is under current structures available largely to the wealthy.

Looking back through the geological record, we see that when the climate changes quickly, the majority of animals and plants cannot migrate or adapt quickly enough. Mass extinctions tend to result.[56] Humans have had the good fortune of evolving during a remarkably stable climate period, the Holocene, which we have now warmed our way out of. Natural climate change would also present challenges (we would still have to shift cities, change crops, confront new diseases), but these can be more easily met by our natural adaptative capacity because they would unfold over a much longer period. By speeding up the rate of climate change, we make the task of adapting difficult, if not impossible. Looking at climate change in this way means the question is really, can we prevent (mitigate) climate change fast enough to keep the impacts small and slow enough for us, and the natural world, to adapt?

Second, seeing the problem in this way focuses our attention on the factors that determine the rate of change. The rate of mitigation of course depends on patterns of consumption and production, technological change, and crucially, the policy choices we make to affect these. For adaptation, policy is also of course important, but the literature focuses our attention on economic resources and governance capacity as key determinants of adaptation ability. To put it bluntly, the richer a society is and the better able it is to channel its resources into farsighted physical and social infrastructure, the more it can adapt. Managing the impacts of climate change will require vast deployment of physical infrastructure, restoration of natural ecosystems, new models of agriculture, and significant movement of not just individuals but whole communities. The rate and difficulty at which all this can be done will depend on material wealth and effective governance but also, scholars contend, more intangible attributes like social capital.[57] A community with a strong sense of solidarity may be able to rehouse populations from low-lying areas without exacerbating social tensions—neighbors will help neighbors. In a polarized community with low social trust or intergroup animosity, such a relocation program could create a major conflict.

Looking at these determinants of the rates of change for mitigation and adaptation reveals a particularly concerning aspect of climate change. Some of the factors that affect the rate of mitigation can change relatively quickly. For example, renewable energy technology has gone from fringe to mainstream in a few decades. Our consumption patterns can rapidly shift at scale (think of the rapid rise and fall of smoking around the world). But the key determinants of the rate of adaptation—wealth, governance capacity, social capital—are slow to change. They grow out of complex historical processes, often shaped by critical junctures and path dependence (see above). We can (indeed, must) strive to eliminate poverty, strengthen governance capacity, and enhance social capital. But research and experience suggest this work is long and nonlinear. The implication is that the fundamental drivers of adaptation may not change as quickly as the climate itself changes. Barring radical change, we will be adapting at a pace determined by the wealth, governance capacity, and social capital presently available.

Worse, to the extent the rate of climate change exceeds the rate at which we can adapt, a new danger arises in the form of a sociopolitical tipping point. The determinants of adaptation—wealth, state capacity, social capital—are not just slow to build up; they are also vulnerable to erosion by the impacts of a changing climate. Natural disasters, agricultural declines, resource conflicts, spreading infectious diseases, mass migration, and other climate impacts all have the potential to destroy wealth, weaken governance capacity, and sap social capital. Because these factors emerge from complex historical processes and because the impacts of climate change are so diverse, we cannot say that climate impacts will always undercut the determinants of adaptation. But certainly, in many cases they will, and it seems probable that on average we can expect a negative effect. This relationship creates a terrifying prospect: could the rate at which we mitigate and adapt fall below the rate at which the impacts of climate change disrupt our ability to adapt? If so, we may reach a point at which we (or at least those people, places, and societies with less wealth, governance capacity, and social capital) can *never* adapt fast enough, losing all ability to control the problem. Should we pass that critical threshold, defined by comparing the determinants of the rates of change of different elements of the problem structure, it is difficult to see a way back. As the 1970s bestseller *Future Shock* put it, long before climate was a matter of political debate, "unless man quickly learns to control the rate of change in his personal affairs as well as society at large, we are doomed to a massive adaptational breakdown."[58]

Finally, we can think about solutions to climate change in terms of achieving the right balance between rates of change. The implication of the previous point is that we must keep the rate of climate change slow enough, through mitigation, to not overwhelm the factors that determine the pace at which we can adapt. We are not seeking to permanently fix the average temperature but rather to slow it back down to a more natural range of rates. If we can do that, then the rate of change becomes essentially imperceptible, falling back within the normal adaptative capacities of human and natural systems. Climate change will not stop, but it would become a "super long problem," measured in millennia rather than decades or centuries, much more commensurate to the timescale at which we have adapted historically. By the same token, building up our ability to adapt by fostering economic development, state capacity, and social capital can help us manage a faster rate of climate change. Defined in this way, solving climate change means mitigating quickly enough to slow it down to a pace at which we can adapt with reasonable comfort.

Dynamic Problem Structure Chains

In the time between cause and effect, many elements of a long problem's structure can change. We can expect these shifts to be driven both by factors at the core of the problem (such as different rates of change, as explored above) and by those otherwise unconnected to it. But whether the shifts are endogenous or exogenous, changing problem structures present challenges to social scientific approaches that seek to hone in on a particular causal factor or mechanism to describe a relationship between elements of interest. The point of any theory is of course to simplify complexity in a way that facilitates understanding. Parsimony is therefore necessary to theorize and examine a given causal relationship. But it also comes with an analytic trade-off. Either the scholar must

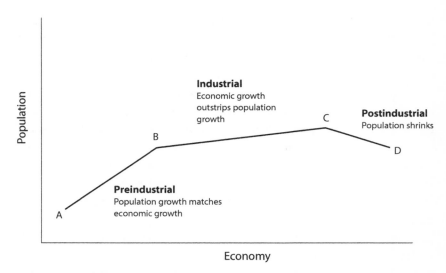

FIGURE 6.1. Stylized relationship between the economy and the population at different periods

assume, perhaps implausibly, that the elements of the problem structure she is analyzing will not change, or she must accept that her explanation faces temporal scope conditions.

Recall the Malthusian example above, rendered in figure 6.1. Malthus's hypothesis that population growth matches economic growth (perhaps) holds in the preindustrial period between A and B; a one-unit increase in the economy leads to commensurate increase in population, and so wealth (e.g., GDP per capita) never changes. But then between B and C, technological changes lead economic growth to rocket ahead of population growth, and wealth accumulates. Later still, at higher levels of economic development, we may actually see a decrease in population over time, as wealthier societies tend to have fewer children. The causal mechanisms driving these inflection points are partially endogenous, reflecting the technological and social change that occurs at higher levels of material wealth but also reflects at least partially exogenous factors like scientific discoveries and changing gender norms.

How social scientists assess the relationship between the population and the economy therefore depends entirely on the period of analysis. Parsimonious explanations that hone in on a core causal relationship may offer the best way to explain a given segment between inflection points, or perhaps a transition from one segment to another, but they do not generalize across segments and would be at best incomplete, at worst misleading, if used to explain the entire period from A to D.

For long problems, then, the analytic value of parsimony diminishes because we need to understand the different causal relationships at play in distinct periods, as well as the ways in which they shift from one to another. A problem's structure is more likely to transition through different phases. Our theories therefore need to go beyond "X relates to Y like this" and should look more like "first X relates to Y like this, then because of change B like this, and finally because of change C like this." A long problem requires a chain explanation, with the different causal mechanisms at work in each period concatenated by (endogenous or exogenous) explanations of change. Only with this additional complexity can we explain a dynamic problem structure chain with sufficient completeness.

That said, a simple list of temporally "local" theories would be equally unsatisfying, with understandings of long problems simply amounting to "one damned thing after another."[59] We also need more general theories that stretch across different phases of a long problem. Ideally, we should think of these approaches as complementary and nested, with more focused theories explaining local contours and more general theories explaining across phases. Seeing theory in this way will require an adjustment in our usual social scientific procedures because one piece of research, or even one scholar, may not be able to capture all of these elements in a cogent and coherent way. Instead, researchers should be looking to build a collective theoretical apparatus—that is, a literature—that covers all aspects of a long problem.

To be clear, even for long problems with dynamic problem structures, it remains worthwhile to focus on individual periods or transitions. Indeed, such targeted analysis is likely necessary to understand the problem at any given point in time. However, such explanations need to be joined together temporally to understand the broader context, which makes their scope conditions transparent.

How can we apply these ideas to climate change? A long approach must theorize what inflection points in the problem structure are possible, and then analyze the political dynamics in the various segments between them, as well as the transitions across them. Recall the various theoretical approaches to climate migitgation posited above: collective action, distributional politics, polycentrism, transition, catalytic cooperation, asset revaluation. If we look across the full span of the climate problem, considering also climate impacts, at which points are these different lenses most helpful?

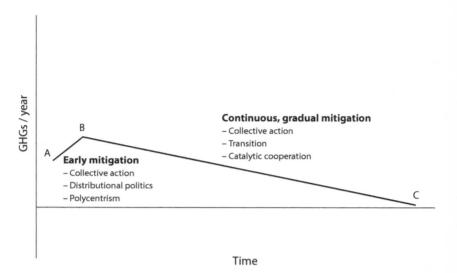

FIGURE 6.2. Scenario 1: Prevention (the road not taken)

Below, I present three dynamic problem structure chains for climate politics from roughly the last decade of the twentieth century through the remainder of the twenty-first century. Each chain outlines a potential scenario; there are countless others.[60]

The first, in figure 6.2, is counterfactual. How would the problem structure of climate change look if strong action had been taken from the time the problem was first identified? Imagine that the early international conferences in the 1990s succeeded in creating sufficient collective action on climate change to drive down global emissions. Credible political commitments from major emitters led to the early development and deployment of new technologies, which, through effective international cooperation and support, diffused globally, following the model of the Montreal Protocol. Wealthy countries moved first, but a combination of newly developed technologies and international financing allowed developing countries to quickly follow, avoiding high-carbon-growth paths. Potential laggards were deterred by credible threats of economic sanctions from a critical mass of countries. With this shared commitment, early action ensured that carbon emissions peaked at a low level, leaving enough atmospheric space for mitigation to proceed gradually over the following several decades. This gradual rate of change softened the distributional impacts of decarbonization, giving interest groups tied to fossil fuels adequate time to adjust and transition. Similarly, climate impacts remain at low levels into the future, allowing adaptation to build up over time. Large-scale carbon removals and geoengineering are not required because reducing emissions generates sufficient mitigation to slow climate change to a bearable speed.

In this counterfactual scenario, the problem structure of climate change remains relatively constant, meaning that the theoretical lenses face fewer temporal scope conditions. Collective action explanations remain central throughout, as collective commitments drive policy and shape expectations. There is also a strong role for transition theories, particularly those that emphasize experimentalism (the development and deployment of new technologies) as well. Distributional politics of course also matter since the scale of the transition requires widespread adjustments, but the gradual pace of transition eases and disperses the costs that negatively affected groups face, making it easier for Pareto-improving bargains and compensatory arrangements to be agreed, such as retraining of coal miners or long-term investments in economic diversification in oil-producing states. Because mitigation largely works, the politics engendered by extreme climate impacts remain muted. Writing three decades after the first UN climate conferences, it is easy to dismiss this scenario as naive. Analytically, however, it provides a useful counterfactual.

The second and third scenarios start from the time of writing, when the window for early mitigation has passed. The same theoretical lenses help explain climate politics over the last three decades, though the outcomes differ. We now know that collective action did not succeed in generating sufficient policy change to allow gradual mitigation. Instead, action has been blocked by economic interest groups that would have been adversely affected, as theories of distributional politics would expect. Now only rapid decarbonization can prevent the worst effects of climate change, though some impacts are already occurring and will get worse. We thus look ahead to two stylized scenarios, one in which that rapid decarbonization occurs, and one in which it does not. Because climate change has not been successfully prevented, its problem structure begins to change.

In the high-mitigation scenario (figure 6.3), transition theories help explain why the pace of mitigation increases from slow to fast. Per the catalytic cooperation model, action by first movers changes the costs and benefits that other actors face, allowing mitigation to accelerate over time. Per the asset revaluation model, climate-vulnerable interest groups mobilize on a large scale as the threat they face from climate change becomes increasingly existential. If these dynamics are sufficient to disrupt fossil fuel lock-in, changing the political balance of

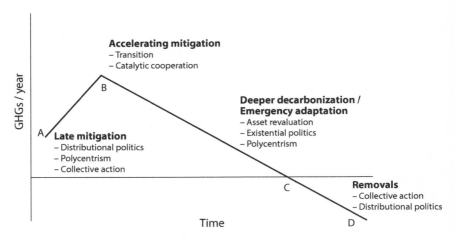

FIGURE 6.3. Scenario 2: Delayed rapid mitigation

power decisively in favor of decarbonization, mitigation may accelerate even more quickly. Holdouts and laggards resist and fight back with increasing intensity and perhaps violence as they face extinction. But ultimately, their political power erodes as the economic value of fossil fuels evaporates, and pro-decarbonization interests are able to either defeat them or buy them out.

However, although decarbonization is ultimately successful, it comes too late to avoid significant impacts. As time goes on, climate change is increasingly perceived as a problem of adaptation. Indeed, climate is seen less and less a discrete problem and instead a pervasive background condition affecting a wide range of related issues at multiple scales—health, food security, conflict, extreme weather events, and so forth—each with its own distinct problem structure. In this sense, the polycentric lens becomes increasingly important analytically. Moreover, because decarbonization efforts were slow to begin, mitigation remains incomplete. To bring atmospheric concentrations of GHGs to safe levels requires not only halting new emissions but also retroactively removing some that have already been emitted. Climate mitigation therefore shifts from a problem of transitioning away from fossil fuels to one of how to remove a vast amount of GHGs from the atmosphere retroactively, potentially including direct air capture (DAC) or burning biomass for energy and capturing the resulting carbon permanently (termed bioenergy with carbon capture and storage, or BECCS).

The problem structure of these kinds of GHG removals is close to that of a pure public good. It is nonrival (if I do it, you are not less able to do it) and nonexcludable (decreasing atmospheric GHG concentrations benefits everyone). And because it will be costly at the scale we will likely require, providing enough removals is likely to be a daunting collective action problem, perhaps bringing the problem structure of climate change full circle. At the same time, different kinds of removal technologies will have sharp distributional consequences. For example, if we rely heavily on biological solutions, those who live in and around forests will be significantly affected. In other words, the problem structure of climate change toward the end of the period, in this scenario, may be a kind of mirror image of its problem structure at the beginning, a collective action problem (though one based around getting actors to contribute to global public good, as opposed to preserving a global common pool resource) interlaced with distributional politics.

In the low-mitigation scenario (figure 6.4), the problem structure is similar to begin with, though the outcome is quite different. Transition theories that examine the conditions under which mitigation accelerates also help explain why it may not. In this scenario, the catalytic effects of first movers are insufficient to generate substantial changes in other actors' preferences; leadership fails to engender much followership. Similarly, although climate-vulnerable interests increasingly mobilize to defend themselves by promoting mitigation, they are unable to rout fossil-based interests quickly or fully. Mitigation may accelerate to some degree, but decarbonization remains incomplete and fiercely contested, generating existential politics but without a clear resolution in favor of pro- or anti-decarbonization interests.

As a consequence, climate impacts metastasize as the world warms past 2°C. These impacts become increasingly disruptive, prompting both short-term crises around, for example, extreme weather, and longer term challenges like the abandonment of low-lying coastal lands or regions prone to excessive heat. As such, challenges intensify and multiply; they, not mitigation, become the primary lens through which actors understand the climate problem. Efforts to reduce emissions and prevent future impacts lose salience in the face of immediate threats. As noted above, the manifold challenges of adaptation do not follow a single problem structure, but as climate impacts deepen, the steps required become more radical and disruptive, such as mass relocation of populations, carrying a larger potential for contestation and conflict. In a similar vein, having lost the luxury of gradual mitigation through emissions reductions and removals, actors may turn to geoengineering solutions like stratospheric aerosol injections, seeking to artificially cool the planet. The problem structure of such approaches could resemble that of mass carbon removal—a collective

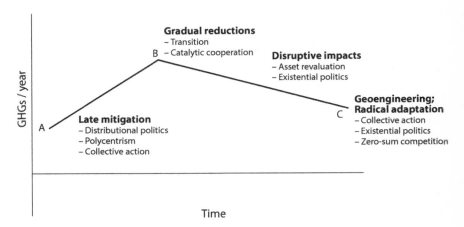

FIGURE 6.4. Scenario 3: Delayed slow mitigation

action problem of how to pool sufficient resources to make a difference. But they may also endanger sharp conflict if, for example, geoengineering interventions succeed in cooling the planet but disrupt local weather patterns. To the extent they do, their problem structure may be defined by a Hobbesian logic of zero-sum political competition between actors, with each aiming to save themselves.

These three (nonexhaustive) scenarios outline different problem structures we may anticipate over the course of climate change, as well as explanations of the transitions between problem structures. These are best summarized in dynamic problem structure chains. Only the early mitigation scenario has essentially the same problem structure throughout. In this (counterfactual) case, a chain explanation is unnecessary. The two delayed mitigation scenarios, in contrast, follow different sequences of problem structures, with a range of theoretical lenses providing different insights at different points.

These short scenarios demonstrate why long problems tend to require more theoretical apparatus to explain—they are prone to change over time. To the extent they do, social scientists need to consider dynamic chains of problem structures and corresponding sequenced packages of explanations. Of course, not all work can or should aim to cover the maximal period in which climate change will be a political challenge. But to the extent our theories zoom in, they should do so in a way that is sensitive to potential temporal scope conditions or risk overgeneralization or myopia.

Empirical Analysis of the Future

Can we study what has not yet happened? The epistemological difficulty of doing so has not prevented many from trying. Ancient humans relied, and many today rely, on various forms of mysticism or folk wisdom to divine the future.[61] But science allows for a more evidence-based approach. As the scientific revolution spread to the social realm in the nineteenth and early twentieth centuries, the notion of future studies emerged. In 1902, none other than H. G. Wells, the writer who did perhaps more than any other to spark the modern genre of science fiction, argued for the scientific study of the future in a famous lecture:

> I believe that the deliberate direction of historical study of and of economic and social study towards the future and an increasing reference, a deliberate courageous reference, to the future in moral and religious discussion, would be enormously stimulating and enormously profitable to our intellectual life.[62]

With the advent of computer-assisted statistics in the middle of the twentieth century, "futurology" took another step forward. Researchers at RAND used new quantitative methods to seek to understand the likelihood of nuclear war and other outcomes of interest, labeling their technologically based approach cybernetics.[63] The approach became highly influential in American military planning. The fantasy of prediction was captured in the seminal science fiction work of Isaac Asimov, whose *Foundation* books center on a social scientist who was able to predict, and therefore influence, the future.

With the advantage of hindsight and with a more developed view of social science, we can chuckle at the hubris of these early futurologists. Prediction, in the sense of knowing what will and will not happen, is clearly impossible in nearly all domains of social life.[64] Even the less ambitious task of assigning conditional probabilities to a range of outcomes regularly fails, no matter how much human effort is deployed, such as in financial markets or intelligence services. To formally evaluate the accuracy of expert judgments, the political scientist Philip Tetlock conducted a series of prediction tournaments in which experts made predictions about various events, which were later evaluated after they happened.[65] The results were unimpressive: "In aggregate, experts edged out the dart-tossing chimp but their margins of victory were narrow."[66] And experts failed to beat either knowledgeable nonexperts, a group Tetlock summarized as "attentive readers of the *New York Times*," as well as mechanistic algorithms that simply extrapolated past trends into the future.[67] Only undergraduate students fared worse at prediction than supposed experts.

The results are sobering but perhaps not surprising. The complexities of predicting social or political outcomes are massive, not least because efforts to understand the future can themselves alter that future by changing the way we act in the present. Social scientists rightly understand that our increasingly sophisticated theories and methods, no matter how powerful they might become, can never transform us into soothsayers or oracles (and we are skeptical of those who pretend otherwise).

Still, H. G. Wells's 1902 insight is not wrong: we do have a need to study the future. Moreover, the long-standing, seemingly universal human desire to have a way of thinking about what comes next—even if it is just animal intestines, tea leaves, or a gut feeling—suggests a deep-rooted impulse to understand the future exists across nearly all human knowledge systems. Looking ahead is something societies do, so as Barbara Adam puts it, "if people move freely in the entirety of the temporal domain, then those who study society need to follow suit."[68] The more long problems we face, extending decades and centuries in front of us, the more that need grows. As Andreas Wenger and colleagues argue, "being policy-relevant means to supply future-related, forward-looking knowledge—a task that does not come easy to a profession that traditionally focuses on the empirical study of the past and present."[69] But even from a purely theoretical perspective, students of social science need to be able to look at those parts of a long problem that have not yet happened to theorize them correctly. But can it be done as part of social science? There are at least two reasons to think so.

First, even though the future itself is not observable, its ingredients are. The things we do individually and collectively today, plus the things we have done in the past, create the reality we experience tomorrow. Adam explains this clearly, arguing that although people in the past may have seen the future as mysterious and inscrutable, a scientific worldview leads to a different interpretation:

> In modern times, we instead see the future as ours to make. In contrast to predecessors, we do not think of the future as already existing or predetermined by supernaturals. Instead, we take for granted that the future is created by us, in and for the present. We see ourselves as producers and managers of an open future, which we shape and negotiate to our plans and intentions. Thus, the future is assumed to be subject to our will, even if it does not always work out the way it was planned.[70]

In other words, in the modern worldview, future outcomes *must* be fashioned from the raw materials—ideas, institutions, behaviors, material resources, and so forth—of the observable present and past. There is nowhere else the future could come from. It is latent in processes we can observe, and this creates an important point of analytic leverage. I cannot predict what you will cook for dinner tonight, but if I know what ingredients and cooking apparatus you have at home and where your tastes lean, I can begin to specify quite a lot.

Second, we do not need perfect certainty about the future to study it scientifically. Knowledge and information of a more uncertain ontological status can also be grist to the social scientific mill—indeed, it already is. We perhaps miss this because contemporary social science prioritizes post hoc explanation as the ultimate goal, creating a very high (and perhaps artificial) threshold for what counts as valid knowledge in the scientific process. Roughly, we seek to show that X causes Y by measuring both X and Y, theorizing the causal mechanisms that link them and testing hypotheses against observed data around the implications of both our theorized mechanisms and alternative explanations. When successful, this process is meant to establish robust evidence that X has caused Y. If we see social science as only this process, then studying future outcomes is nonsensical because the data to describe X and Y do not yet exist and so the testing cannot be done. At most, social scientists may posit that if the conditions around this causal link are sufficiently general, we may expect X to cause Y again should sufficiently similar conditions emerge in the future. This final step is usually added as an afterthought in an article's conclusion. For studying long problems that extend into the future, however, we need to understand how a causal process may develop going forward.

Although much has been written on this topic, Gunther Hellmann offers a usefully cogent critique of the narrow, exclusively backward-looking interpretation of the scientific process.[71] Following Dewey, he advocates for a pragmatic approach, questioning whether the "quest for certainty" ultimately limits knowledge. Instead of looking only for regular laws, he suggests that a better goal for social science is to establish tendencies—X may or may not *cause* Y, but we can at least say that X *tends toward* Y. Indeed, this is what much of empirical social science looks like in practice, given measurement limitations, multicausality, and the general difficulty of establishing laboratory-like experimental conditions. This wider understanding opens space for studying the future because we need not argue that X will cause Y (essentially, a prediction about the outcome) but rather that X will push toward Y, a more modest but equally scientific claim about tendencies. As Tetlock notes, theories can be robust and valid without also being required to be 100 percent predictive; "it is possible to be right about underlying causal drivers but unlucky on the fluky details."[72]

If we set the threshold for valid knowledge at a level of certainty only obtainable by post hoc causal testing, we will struggle to get much analytic traction on problems that extend into the future. As Toffler put it in *Future Shock*, "the inability to speak with precision and certainty about the future, however, is no excuse for silence."[73] In short, social scientists should become much more comfortable asking not just why and how an outcome can be explained but also what if and how might future outcomes manifest.

Because the elements that create future outcomes are observable now and because we do not need full certainty about future outcomes to study them, the tools of positivist, empirical social science can be brought to bear on the future. But before discussing how to do so, it is critical to emphasize one other element of the scientific process that is essential for effective study of the future: skepticism. As the above examples of futurology gone wrong remind us, the hubris of the analyst is a particular danger in this area. Rigorously identifying possibilities and tendencies is important, but we mislead ourselves if we conflate it with prediction.[74] Perhaps ironically, Tetlock finds that individuals who are more uncertain and tentative tend to make better predictions than those who are very confident.[75] Studying the future is qualitatively different than studying the past because something unexpected always may occur. Indeed, unexpected things occur relatively frequently. This pervasive condition of uncertainty puts important boundaries on the kinds of claims we can hope to make by studying the future because we can never be sure. But this limitation is not a reason to abandon the scientific method when looking forward. Indeed, under conditions of uncertainty, social science is an effective framework for examining the tendencies and possibilities we expect to unfold forward in time because it posits claims, examines them with evidence, and revises them iteratively. If done with humility, this process can accumulate useful knowledge about the future without ignoring the fundamental uncertainty that applies to all future events.

The question, then, is how. Studying the future empirically will require innovation in the methods and to some degree the epistemology of social science. There may be some social scientists who still model their thinking on the (idealized) approach of experimental physics: derive a hypothesis from theory, test it experimentally, and if it survives, an ironclad law of the universe is added to the body of theory until disproven. Many social scientists today would probably subscribe to a more nuanced, Bayesian-influenced version of this approach: identify prior expectations, identify patterns that match or do not match those expectations, update your expectations accordingly, and repeat continuously, zeroing in on central tendencies over time. The difficulty with the first approach is that its logic assumes there are universal laws to be uncovered. As Heikki Patomäki argues, social scientists err when we "apply deductivist ideas and methodical tools that are only suitable for use in closed systems of directly observable phenomena. Outside astronomy, these systems are only to be found in the laboratories of classical, Newtonian physicists. Outside artificially created existential and causal closures, there are only open systems."[76] The Bayesian approach improves on this rigidity by allowing findings (tendencies) to update incrementally over time. This works well when there are many observations and frequent updating is possible but less well when there are large single events, shocks, or trends (such as climate change) that provoke discontinuities. Both are equally backward looking.

As Steven Bernstein and colleagues have written, "God gave physics the easy problems."[77] For social scientists looking at long problems that extend into the future, we need an epistemology that does not assume the existence of physics-like laws that generalize across all future cases (lest we stray into fallacious prediction). Nor can we only generate expectations by looking at only past data. Bernstein et al. posit evolutionary biology as a more useful analogy for social science than physics. After all, its building blocks and tendencies can be identified, and so the future development of organisms can be understood if not predicted.

To be more future-oriented, social science should not (indeed, cannot) give up the idea of identifying regular tendencies by studying present and past phenomena and organizing the causal mechanisms into theories that explain why outcomes are the way they are. But we also need tools to combine these tendencies, derived from theory and observation, into sets of contingent, conjunctural possibilities about future outcomes. Indeed, doing so in a rigorous and consistent way can help build our theories further. Three methods may be of particular importance: computational models like integrated assessment models or epidemiological models, agent-based models and simulations, and scenarios.

As noted above, integrated assessment models are essentially sets of algorithms that relate social and economic factors like the rate of economic growth, the degree of technological change, or the price of carbon implied by different policy regimes, to earth system outcomes like changes in average global temperatures. IAMs draw on multiple disciplines, but their intellectual origins are closest to macroeconomic models that seek to understand trends in economic growth as well as energy system models that provide tools for measuring and matching the demand for and supply of energy. IAMs seek to join together these distinct modules into a combined model that can take a range of observed data (e.g., GDP, current population levels, current land use patterns) and assumptions about how different subsystems work (e.g., the economy, the energy system, the land system, the climate system) and produce integrated estimates for outcomes of interest such as carbon emissions or economic growth. Properly understood, IAMs are not ways to predict what will happen across all these macrosystems but rather tools for asking what-if questions like, all else equal, "How much more temperature change will result if we delay reductions by another decade?" or "What rate of technological change would be needed to achieve a given level of economic growth and emissions reduction simultaneously?"

A significant limitation of IAMs to date is their failure to incorporate key social scientific insights, though this is beginning to change. Students of politics and other social scientists can doubtless improve IAMs by injecting them with some of our core insights. This will improve all disciplines' ability to ask meaningful what-if questions about the future, which is useful for understanding a long problem like climate change. But in making this contribution to the broader field of knowledge, political scientists could also gain a rewarding new area of inquiry for our own discipline, developing a tool to ask what-if questions about the future of politics, with political variables either as explanatory or outcome variables. What effect could we expect on the climate if the rollback of democracy continues? How might temperature change affect superpower competition? Such substantive and interesting questions should not be left to the speculative conclusion sections of journal articles when methods like IAMs can begin to give us some analytic traction on them.

A similar but distinct way of modeling the future, much more familiar to social scientists, is the use of agent-based models. These tools are also useful for asking what-if questions by examining different assumptions. But instead of families of algorithms that seek to model the economy, the climate system, and so forth, they rely on stochastic interactions between different agents—individuals, firms, countries, and so forth. Political scientists have used such techniques to understand the prospects for cooperation on climate change,[78] and economists have used them to understand prospects for decarbonization among firms.[79] Some scholars have even combined agent-based models with IAM techniques to better capture the complexity of social interactions.[80]

Another way of generating knowledge about future tendencies is through simulation exercises or experiments not with computer algorithms but with humans. Militaries regularly conduct war games with the aim of generating battlefield options and possibilities that may not emerge from simple desk work. Not limited to military applications, these kinds of test runs provide a useful heuristic for understanding how actors may behave when "parameterized" with certain conditions. More scientifically, researchers run laboratory experiments in which groups of people interact. By controlling the conditions that shape the interaction, researchers are able to examine how different factors can hypothetically shape outcomes. For example, Scott Barrett and Astrid Dannenberg study cooperation between students representing countries. By giving different groups more or less certainty around the threshold at which catastrophic climate change takes place, they show that uncertainty can provoke more challenging negotiations.[81]

Finally, Bernstein et al. advocate for the use of scenario-style approaches in social science, an idea Levin et al. also advance in the form of "applied forward reasoning."[82] Scenarios "allow researchers to combine general knowledge of politics with expert knowledge of individual actors and situations, to build in context, complexity, variation and uncertainty in the form of multiple narrative with numerous branching points, and to revise their expectations as events unfold. Repeated iterations of this process can reasonably be expected to improve the quality of our general knowledge of international relations, our ability to track specific developments and the outcomes that result, and our capacity to address the problems that these evolutionary tracks create."[83]

There are many methodologies for constructing scenarios, not all of which will be of use to social scientists. Broadly, Bernstein et al. suggest seven elements to a productive scenario:

1. Identify driving forces, those causal mechanisms most likely to affect the outcomes of interest.
2. Specify predetermined elements, those factors that have already occurred or are very unlikely to change.
3. Identify critical uncertainties. Are there significant causal factors whose influence or value is difficult to specify? What different possibilities might arise should that factor go one way or another?
4. Develop scenarios with clear "plotlines," causal chains that show what factors matter at different times, and how one set of circumstances leads to another.
5. Extract early indicators for each scenario, generating ex ante observable implications for different pathways.
6. Consider the implications of each scenario; what would we expect if the outcome of interest developed as the scenario describes?
7. Develop "wild cards," phenomena that are difficult to foresee but that could have large impacts on the outcome being studied.

Although we tend not to use the language of scenarios, many of these steps should be very familiar to social scientists. First, identifying causal mechanisms is the goal of social scientific theories, the normal work of generating hypotheses. Second, understanding predetermined elements—for example, structural conditions or institutions—is also a key part of developing theoretical expectations, as is third, identifying critical uncertainties. The latter focuses our attention on what the key variables of interest should be. The fourth element, developing "plotlines," is commonly used in process-tracing research or in the development of analytic narratives. Specifying observable implications, the fifth element, is a necessary step for all theory testing, though in the scenario method, these remain expectations to be tested in the future, not to be applied to past data. The last two elements, considering implications and wild cards, is not typically part of the social science process but are useful heuristics for theory-building. They force the social scientist to ask, "If X were to happen, what would my theory expect?"

Given the many similarities between the process of scenario analysis and the existing work of social science, it should not be controversial to include more future-looking analysis in the social science tool kit. As Bernstein et al. argue, "Using scenarios as a research method, the goals of research expand to include not only the development of better explanations, but also identification of points of intervention, ongoing revisions of scenarios as events unfold, and the consideration and re-evaluation of salient causal pathways."[84] This can only help social scientists better understand the world and especially those parts of long problems that have not yet occurred.

IAMs, agent-based models, and scenarios all provide concrete methods for social scientists to study the future empirically. But when we do so, we need to take an extra dose of the humility and self-reflection that good science requires. When we study the past, there is no chance we can influence it (though our interpretations of it may of course affect the present and future). But to study the future is to influence it, potentially decisively, and so scientists must act accordingly. This is not new. Climate modelers, epidemiologists, or economists who engage in study of future events all must make scientific claims with the knowledge that they could influence outcomes. In political science as well, students of election polling or war often make claims about which candidate is likely to prevail or when the conditions for conflict heighten. This potential for influence creates an additional duty to be fastidious in highlighting the assumptions, uncertainties, and limits of our analyses. As a discipline, we need strong professional norms to hold ourselves to account for these high standards. As Adam notes, technical experts today are, in some ways, not so different from the priests of old who sought to divine God's will to instruct people how to behave. The methods have become more scientific and the authority is perhaps diminished, but the responsibility is no less important. Ultimately, understanding the future is a key part of governing it, to which the final chapter of this book now turns.

7
Governing Time

From climate change to technological development, to demographic shifts, long problems are growing. As the problems we face expand across time, our tools for organizing human societies—politics and governance—must similarly adapt to manage the challenging political economy long problems present, marked by shadow interests, institutional lag, and centrally, the early action paradox.

Together, these features make long-term governance difficult but not impossible. The previous chapters have shown the particular conditions under which and specific mechanisms through which they can be tackled. These include informational mechanisms, foresight techniques, experimentalism, catalytic interventions, representatives of and trustees for future generations, rule changes that extend actors' time horizons, capacity-building and education, citizens' assemblies and other deliberative mechanisms, advocacy, goal-setting, reflexive updating of institutions, and triggers and reserves (table 7.1). Examples of these tools can already be found across all kinds of countries, democratic and autocratic, rich and poor, as well as in regional and global governance, though they are predictably most common, developed, and effective in states with more governance capacity.

Applying a political lens to long problems reveals that for any of these approaches to bite, the critical question is always, do conditions exist to allow these tools to actually change incentives now? Do the ways actors develop certain interests, the ways they build, wield, and contest power to advance those interests, and the ways institutions structure their interactions change? Although the answer is not always yes, these tools of long-term governance can often have at least some grip on the present. This understanding offers a qualified optimism that although long problems are hard problems, human societies possess tools and agency to tackle them.

Table 7.1. Categories of solutions to the challenges of long problems and their limits

Challenge	Categories of solutions	Conditions and constraints
Early action paradox	**Information and foresight** to make the future known and salient. Example: Environmental or fiscal impact assessments, foresight. **Experimentalist governance** to tackle uncertainty. Example: Montreal Protocol. **Catalytic strategies and institutions** to seed and scale actions and processes that can shift actors' preferences over time. Example: Paris Agreement.	Must mobilize actors in the present who have long-term interests; effects depend on the political power those actors possess. Requires political support for goal and ultimately coercion; learning and diffusion must be possible. Requires channels (material, information, normative, or political economy) through which action by first movers shifts incentives.
Shadow interests	**Representatives for future interests** (in general or for a specific issue) Example: UK Climate Change Committee. **Trustees** with mandates and power to act for long-term interests. Example: Court rulings requiring action to meet climate targets. **Shift actors' time horizons** through rule changes, capacity building, activism, and other processes. Example: Climate assemblies.	Impact depends on ability of representatives to use their agency and the political power of aligned actors in the present. Present actors must have incentives to create/maintain trustees, which must be able to resist short-term pressures. Slow processes.
Institutional lag	**Long-term goals** to make action durable. Example: Sustainable Development Goals. **Reflexive governance** to update policies and institutions regularly. Example: Review and sunset clauses. **Triggers and reserves** to enable updating and ensure resilience. Example: Sovereign wealth funds.	Must have ways to influence actors' present incentives and remain updatable; competing goals may stymie action. May undermine institutional durability; requires political conditions that favor ongoing bargaining, e.g., trust. Difficult to foresee future shocks or institutionalize long-term restraint on use of reserves.

Still, it is hard to examine the political economy of long problems and conclude that our political systems, as they currently exist, are well set up to address them. Climate change, for example, is now far from the realm of first-best outcomes, though significant scope still remains to avoid the worst scenarios. To make progress, we need to take the political challenges of long problems seriously. This concluding chapter therefore proposes a bold institutional agenda for addressing climate change and other long problems. Governments should

- adopt informational tools like independent assessment entities to better understand the future and to make it salient, including foresight techniques;
- delegate power to trustees mandated to consider long-term interests, including both existing bodies like central banks and courts and new ones with climate-specific mandates and legal powers;
- expand the role of participatory deliberation by citizens and social groups in the climate policymaking process, including via permanent climate assemblies with real power;
- create more robust and complete goal-setting and forward planning processes around climate outcomes;
- weave updating and trigger mechanisms throughout climate policy processes, including through regularly updated pathways toward long-term goals;
- create reserve funds to invest in mitigation, adaptation, and a just transition; and
- deploy experimentalist and catalytic strategies to shift actors' preferences over time.

Although such reforms come with tensions and trade-offs, they offer a lifeline to help our political systems rise to the challenge that climate and other long problems pose. These proposals are perhaps far from our present reality. However, the logic of long problems invites us to think about feasibility as a dynamic process. Developing long-term governance is perhaps akin to changes as big as the shift to representative democracy in the nineteenth century, the rise of multilateralism and human rights in the twentieth century, or the development of independent judiciaries across this period. Such ideas were mere fantasies of thinkers and activists for decades or centuries. Their development was slow, haphazard, and nonlinear, and they remain incomplete and contested. And yet today, institutions like elected legislatures, independent courts, international organizations of sovereign states, and formally enforced human rights are quotidian realities of political life. Our successors may look back and see long-term governance following a similar path. As the great observer of democracy political scientist Robert Dahl noted in 1989, "Whatever form it takes, the democracy of our successors will not and cannot be the democracy of our predecessors. Nor should it be. For the limits and possibilities of democracy in a world we can already dimly foresee are certain to be radically unlike the limits and possibilities of democracy in any previous time or place."[1]

How can we get there? After laying out an institutional agenda for addressing long problems, the chapter considers the potential of climate change specifically to contribute to the institutional evolution we need. I argue there is no inevitable push from a long problem like climate change toward long-term institutions that can help address it. Indeed, in many ways, the opposite is true: the impacts of climate change threaten to make our politics even more reactive and short term. What the challenge of climate change does offer, however, is the *potential* to serve as a catalyst for longer term governance. The choice to use it as such, or not, is ours.

Finally, the book ends by considering what might be different if human societies become, over time, more adept at governing across time. What could governing the future look like? In the introduction, I cited John Vickers's potent image from his 1970 *Freefall* of human society as someone who has jumped off a tall building and is just beginning to realize the ground is fast approaching. The book concludes by asking, what if we learn to fly?

An Institutional Agenda on Climate Change

It is impossible to attend any UN climate change conference and not be inundated with policy solutions like carbon pricing, feed in tariffs, renewable portfolio standards, blended finance, and so forth. International organizations, nongovernmental organizations, researchers, and others devise and collect these and other policy ideas in countless panel discussions and documents. Regular IPCC reports contain hundreds of examples of policies through which governments can and should address the climate crisis. The Global Stocktake, a review of collective progress toward the 2015 Paris Agreement's goals that takes place every five years, has become one of many processes through which such ideas are actively put to governments for consideration and, ideally, uptake in their NDCs. If reports and roundtables on solutions were sufficient to halt climate change, we would have solved the problem long ago.

The challenge of course is not just to design smart policies but to marry them to effective politics. Here, many observers throw up their hands and bemoan the lack of "political will"—a magical balm that removes all venal impediments to efficient, socially beneficial policies. But for the political scientist, this is not the end but rather the start of the intellectual challenge. How and under what conditions can political power be built to advance at least some of the solutions proffered? To get the outcomes they seek, policymakers, political entrepreneurs, and advocates need to think about not just policy solutions but also "politics solutions." At heart, this means building and using political power.

Many are doing so, now more than ever before. As chapter 4 discussed, climate change is increasingly a subject of debate and contestation at the heart of politics around the world. At the grassroots level, activists and advocates, led by schoolchildren, have pushed the issue onto the public agenda, not just in governments but also in the private sector, educational institutions, community organizations, and across society. Political parties in many countries increasingly orient around the topic (either for or against) and political projects like the US Green New Deal or the EU's Green Deal, to cite some prominent contemporary examples, make it a central organizing principle for political agendas. At the same time, populist and right-wing parties increasingly put opposition to climate policy, especially where it creates immediate, salient costs, at the center of their platforms. Elections have been won or lost on climate stances. In the international sphere, climate is a regular feature of G7 and G20 meetings, the UN climate negotiations attract an enormous amount of media and political attention (including regular appearances by leaders of major countries) relative to their limited decision-making power, and green industrial policy has become a major subject of economic statecraft and geopolitical competition.

As both the impacts of climate change and decarbonization intensify and grow existential for larger and larger swaths of society, political contestation around climate will continue to increase.[2] This politicization carries the potential, certainly not guaranteed, to generate the conditions under which effective policies can be adopted at scale, should pro-climate interests win and hold political power. In this context, it is vital to use advocacy, persuasion, organization, and all the other tools of politics to ensure that political will is not assumed or wished for but rather purposefully built.

This work is urgent, necessary, and also already too late to achieve first-best outcomes (though we are still very much able to avoid the least desirable pathways) in large part because the rules of the political game are stacked against effectively addressing a long problem like climate change. The early action paradox, shadow interests, and institutional lag have held us back. If addressing climate change requires generating the political conditions that can lead to the adoption of smart policies, then it follows that we should work, to the extent possible, to make political conditions more favorable. How can we tilt the rules of the political game?

The primary channel is by changing political institutions. As discussed in chapter 2, institutions help define how power is won, who holds it, and how it can be wielded. Too often, these institutions advantage short-term interests over long-term ones. But the ideas discussed in the preceding chapters show a number of ways to correct that imbalance. Political actors who want to address climate and other long problems in a substantive way should put these kinds of deeper reforms at the heart of their strategy, or they will continue to find that political will for the needed policies keeps falling short. Debates on climate change have primarily focused on what the policy agenda should be. But to get good policies, we need better politics. One of the primary ways to do that is to reform institutions.

Drawing on the ideas in the proceeding chapters, an institutional agenda for climate could incorporate a range of changes to govern more effectively for the long term.

First, as a starting point, every government should give itself the informational tools to better understand the future and to make it salient. These should include some sort of at least partially independent body that studies climate mitigation and adaptation in the national context and issues regular reports to which the government is required to respond, analogous to what the IPCC does internationally. Procedures to study and consider the future should also be worked into the normal processes around legislation and executive and administrative rule-making such that the longer term climate impacts of any major policy decision are brought into the center of the decision-making process. Two particularly powerful ways to do this are to create an independent assessment entity that could rate proposed policies for their impacts on the climate, like we see in the fiscal realm, and to mandate a certain rigorous discount rate or "social cost of carbon" be used in the cost-benefit analyses that typically feed into policymaking. Critically, these forward-looking tools should consider both mitigation and adaptation.

Moreover, these kinds of informational tools should include not only technocratic assessments and forecasts but also more wide-ranging foresight practices that bring in nontechnical or scientific information and engage a wide range of actors beyond government.[3] More open-ended approaches help avoid the pitfalls of excessive technocracy, such as groupthink or tunnel vision, and serve to integrate a wider array of potential future interests. For a cross-cutting long problem like climate change, which affects, for example, the natural environment, human health, the economy, security, social behavior, and other outcomes in complex ways, this wider lens is essential.

Second, governments should delegate additional power to trustees mandated to consider long-term interests. Existing trustee institutions like central banks and courts could be given explicit climate mandates. Some of this can be done very quickly. Political leaders can choose central bankers who credibly commit to incorporate long-term climate considerations into their oversight of the economy. Legislatures can pass laws that explicitly give central banks this mandate. Similarly, executives and legislatures (which in many jurisdictions are responsible for selecting judges) can seek out jurists who evince a strong understanding of the law's duties toward future generations. More powerfully, governments could amend their constitutions to provide explicit consideration of future generations, which courts could then help to operationalize.

Alongside steering existing trustee institutions to pursue climate goals, governments can create new trustees with explicit mandates to protect long-term climate interests. For example, climate change committees could be created that provide not just information and advice, but set binding targets that governments are required to meet under law. Carbon prices could be set by an independent technocratic authority just as monetary supply is fixed by independent central banks. Transition councils could be created in key sectors like steel, cement, or aviation with a legally binding mandate to reach net zero and regulatory powers to engage industry in an experimentalist process of getting there.

Third, alongside these technocratic tools and independent entities insulated from short-term pressures, governments should substantially expand the role of participatory deliberation by citizens in the policymaking process. Politicians at all levels of government can create climate assemblies and give them a degree of real decision-making power. Such assemblies can benefit from and contribute to the informational mechanisms described above. Information-providing bodies can ensure that citizen deliberations unfold with a sophisticated understanding of future trends and the likely long-term effects of different actions, but technocrats and trustees should also be responsive to the priorities and concerns that emerge from participatory and deliberative processes. In particular, these outcomes can serve as important reference points also for trustees charged with making decisions in the interests of future generations. Foresight practices may be particularly helpful for integrating ideas across technocratic information providers, trustees, and citizen assemblies and also for structuring engagement between these actors and traditional political actors like politicians and bureaucrats.

Fourth, governments should create more robust and complete goal-setting and forward-planning processes around the climate outcomes they have targeted. Nearly all national governments in the world now have a long-term net-zero target and an NDC with interim targets. Many regional and local governments have set similar targets. Some of these include clear plans with concrete steps for achieving them in different sectors. But the majority lack significant detail. For example, they might set an overall emissions target, but not break it into specific subtargets for different sectors. Climate impacts are similarly neglected. Shockingly, many governments lack specific adaptation goals or detailed plans for achieving them, even as climate impacts mount in the present.

Moreover, where mitigation or adaptation plans do exist, they are often highly technical with little input or buy-in from individual politicians, political parties, interest groups, or citizens. They tend to be written by environmental bureaucrats (and often, especially in developing countries, with much of the technocratic work done by international organizations or external consultants) and therefore disregarded by finance, planning, transport, building, agricultural, or industrial ministries, which is where authority and resources tend to sit. Without sufficient political input from interest groups and society, overall plans and goals likely lack important information and are less effective as tools for converging actors' expectations.

As such, many of these goals and plans are paper tigers. Without institutional and political weight, the goals and targets cannot acquire the salience and credibility needed to steer policy over the long term. Actors can reasonably expect them to be replaced or ignored should they interfere with the interests of politically stronger actors.

Recognizing this, governments need to create goal-setting processes that have real heft. For example, a legislature could set a national mitigation and adaptation goal in law and mandate an iterative planning processes around it. Incorporating informational mechanisms, trustees, and participatory deliberation into the goal-setting process, alongside politicians and bureaucrats, can help give goal-setting weight and make plans more effective by involving all relevant decision-makers.

Fifth, governments should weave updating and trigger mechanisms throughout climate policy processes. For example, goals and plans at all levels should have regular opportunities for revision. Some of these could be planned (e.g., review plans and targets every five years, in line with the Paris Agreement) while others could be tied to specific triggers (e.g., review plans if emissions do not fall, if there is a catastrophic climate impact, etc.). Additionally, nonclimate-specific policies (e.g., urban planning rules) could be retroactively reviewed with a climate lens.

Sixth, governments should create an array of reserve funds with explicit, long-term mandates to invest in mitigation and adaptation. They should also create trustee-like governance arrangements around these funds to ensure they remain insulated from short-term pressures. A first step, already in progress, is to give existing capital pools, such as those held in multilateral development banks or sovereign wealth funds, an explicit mandate to address climate change. Many governments are also capitalizing various forms of new "green banks" that can grow over time. Turning sovereign wealth funds into climate transformation funds could be particularly promising.

Seventh, this institutional constellation should be used to empower policymakers to develop more experimentalist and catalytic interventions. Broadly shared goals with credible backing from political authorities create the potential for the coercive policies needed to backstop experimentalist governance. Similarly, within the framework described above, catalytic processes would work to build increasing political support for climate goals, iteratively harvesting updates in actors' preferences into new and stronger coalitions for action on climate change, which then support new and stronger policies.

The exact design and mix of these kinds of institutional forms will of course vary from jurisdiction to jurisdiction. Countries have what political scientist Navroz Dubash calls different "varieties of climate governance"; one size does not fit all.[4] Nonetheless, every country will need several reforms like the kind described above, tailored to its particular needs, to address long problems like climate change.

Together, these kinds of institutional reforms could, over time, significantly alter the politics of climate change. Obstructionism by short-term interests would be progressively eroded as longer term interests came to weigh more heavily on present decision-making. Such changes would not come soon enough to achieve first-best outcomes, which have already escaped us. But the sooner such changes are in place, altering the rules of the political game around climate politics, the more likely our societies will be to avoid the worst that climate change can bring. As such, they are critical to support societies through the coming century of intensifying climate impacts. Climate advocates, political entrepreneurs, and policymakers must therefore target not just the policy outcomes that respond to the climate challenge but also the institutional infrastructure that shapes the political conditions from which strong policy can emerge.

To be sure, the reforms listed above are not the only ones needed. Although not the focus of this book, an obvious corollary of putting in place pro-long-term institutions is to get rid of anti-long-term ones. For the particular challenge of climate change in which opposition from incumbent industries and their consumers in the present is the primary block to more aggressive action, institutional reforms that reduce the power of concentrated interest groups on the policymaking process would be an enormous boon. For example, reducing the ability of corporations to influence the political process through campaign contributions ensures that powerful, concentrated industries will have less sway over decision-making. Similarly, in many federal jurisdictions, malapportionment (where legislative districts weight sparsely populated rural areas over densely populated urban zones) tilts the balance of power to fossil fuel–related interests.[5] Beyond government, requiring private sector companies to incorporate climate goals into their businesses models—for example, by requiring disclosure of net-zero transition plans—can substantially reshape firms' preferences and therefore the incentives policymakers face. In places where a regime's political economy is fundamentally tied to fossil fuels—for example, so-called petrostates—a fundamental overhaul of political institutions may be needed.

In any given jurisdiction, there will likely be a range of these kinds of incumbency-protecting institutions that will need to be altered for climate policy to have a better chance of gaining political support. An institutional agenda on climate change should therefore aim to correct them alongside putting in place the long-term-oriented institutions described above.

To the extent they are adopted and effective, long-term institutions will recalibrate who holds and wields power. For this reason, adding institutions for long-term governance to our political systems will create new tensions and trade-offs, both among different long-term tools and between these new additions and conventional political institutions. It is critical to be clear-eyed about these fault lines.

First, if long-term interests are empowered, are present interests disempowered? To some extent, they must be. Because power is by definition relative, weighting one set of interests more heavily (e.g., long-term ones) necessarily means that others will have relatively less discretion and power. But although some have argued that more long-termism means less democracy, however defined, the institutional agenda outlined above is hardly unidirectional. Some of its elements, like assemblies, will give citizens and social groups greater weight in decision-making. Others, like technocratic bureaucracies that provide information or trustees that explicitly act on behalf of future generations will instead play a countermajoritarian function, putting guardrails around the exercise of power by legislatures or executives. Even if the net effect is to enhance what many observers would consider to be core goals of good governance, such as procedural representation of and substantive protection for those affected by decision-making, there will be political winners and losers.[6]

Here, it is important to recall that there are many examples where these kinds of countervailing constraints on the exercise of power help societies follow more legitimate decision-making processes that achieve better outcomes. Consider, for example, constitutional protections for minorities that hold back the tyranny of the majority, regular elections that checks leaders' tenure, judicial review that holds politicians to certain principles, independent bureaucracies that implement policies without fear or favor, or international bodies that hold states accountable for their commitments. Counterintuitively, constraints on actors' ability to do what they want at any given moment can often help societies achieve broader, deeper goals and preserve important principles like liberty and equality.

Second, because knowledge of the future must be imperfect, strengthening long-term institutions means we will also sometimes get things wrong. Our limited knowledge may lead us to take early action that hindsight will later show to be wrongheaded. For example, we might support a less efficient technology which then gets locked in. The use of biofuels as a substitute for gasoline is one such case. Governments spent billions subsidizing ethanol as a substitute for fossil fuels, reinforcing powerful vested interests in the agricultural sector. But biofuels' overall impact on emissions has hardly been transformative, and it is now clear that these investments continue to distract from and delay the mass electrification of transport, which is the current consensus view on the best way forward (though this too may shift in the future). Alternatively, we might take lengthy steps to adapt to a threat that does not materialize, like the feared Y2K bug at the turn of the millennium. To the extent we face long problems, such inefficiencies will be well worth paying, but they will still bite.

Third, an additional challenge will be blending adaptability with durability, which as chapter 5 discussed, can be managed but never eliminated. Climate politics caries an inherent contradiction in that climate policy needs to exhibit both stability to achieve long-term objectives and disruption to effect a radical transition.[7] If we put too much power in relatively fixed goals and well-insulated trustees, we may find that over time, political institutions work to defend the status quo and cannot keep pace with an evolving problem. But by the same token, we cannot rely on more immediately responsive institutions to act in the long-term interest. Update mechanisms may allow short-termism to creep back in. As a rule of thumb, we can institutionalize goal-setting, updating procedures, and triggers with different time periods and thresholds for first- and second-order institutions to help us navigate this tension. But we should still expect our institutions to sometimes be too responsive and sometimes too unresponsive.

Finally, it is important to recognize that state capacity is a crucial enabling condition for any institutional agenda on climate change. Making the proposed tools work and overcoming the challenges they bring requires government institutions with the human and material resources to deliver complex technical and political functions. Countries with weak state capacity will struggle, for example, to set long-term goals and plans with sufficient technical sophistication, to run experimentalist governance processes that require innovation and learning, to create meaningful citizen deliberation, or to create trustees with the ability to accurately promote the interests of future generations. For these reasons, governments should make building state capacity across the world an urgent strategic priority, as chapter 4 argued. At the same time, however, a focus on long-term governance will require at least a partial rethinking of how state capacity is defined. For example, governmental systems that lack technical planning processes but do a good job of engaging stakeholders in participatory deliberation on future outcomes may fare better against long problems than those that have a lot of resources to write rigorous reports that are never read.

As these points show, there is no free lunch in politics. An institutional agenda on climate is critical for creating a political system that is fit for purpose. It also comes with trade-offs, tensions, and constraints. The impetus for this transformation lies not in ignoring them but in understanding just how costly the short-termist status quo is in comparison. In a world of long problems, governance must adapt. Seeing the costs and challenges of doing makes the challenge less easy but not less urgent.

How Do We Get There: Climate Change and Institutional Change

If we are serious about addressing climate and other long problems, we need to advocate not just for good policies but for changing the political rules in a way that promotes the long term along the lines of the institutional agenda proposed above.[8] What are the prospects for making such reforms happen?

As discussed in the proceeding chapters, we should not necessarily rely on institutions to automatically update to address long problems. The climate case shows this well. Climate change gives us a good reason to adapt our institutions to govern better over time. But the effects of climate change on our political institutions will not necessarily lead us naturally to that result. Indeed, the increasingly existential nature of climate politics is likely to shift governments' focus from prevention to reaction, shortening time horizons.[9]

As we push the climate system to further extremes, the costs of climate change will become much more intense and widespread. Not just small islands but whole coastal regions risk inundation. Droughts will cut off water supplies from hundreds of millions of subsistence farmers as well as those who feed global supply chains. Deadly heat will render whole regions uninhabitable or radically dependent on cooling technology. Under these conditions, climate politics will not just be a question of who gets what, when, and how. Rather, climate politics will become a question of who gets to survive.

At the same time, the advance of decarbonization will pose a similar existential threat to companies, workers, regions, and regimes whose economic survival is linked to fossil fuels. Already, hundreds of coal plants and mines have shuttered across the world, taking investments, jobs, and pensions with them. Oil and gas companies may follow coal, putting enormous pressure on countries and political regimes based on the exploitation of these resources.

In other words, the advance of both climate change and decarbonization efforts will not just change the distribution of resources; it will threaten the very existence of large swaths of the global economy and population. How can we expect political leaders and societies to react?

In the face of urgent survival needs, it may be substantially more difficult to invest political effort and resources in preventing further climate change by reducing emissions. Instead, governments will face increasing, and in some cases overwhelming, pressure to limit the harm climate change and decarbonization are causing in the short term. Imagine you are the mayor of a Middle Eastern city in which the nighttime temperature has been over 50°C for the last week. Will you spend the city budget on climate-saving electric cars or climate-destroying air-conditioners?

Broadly, there are four strategies we can take to counter climate change. We can mitigate it by reducing emissions. We can adapt to it by taking steps like building seawalls or developing drought-resistant crops. We can compensate those who are hurt by its effects to reduce suffering. Or we can, perhaps, deploy technologies to suck carbon from the atmosphere or even directly shield the planet from the sun. We already are doing a combination of these things, though to date, we have focused mainly on mitigation. But as climate politics get existential, political incentives may shift to more defensive approaches.

Indeed, the proceeding chapters have shown that we are already seeing a growing emphasis on such strategies. When the countries of the world pledged, in the 1992 UNFCCC, to tackle the climate problem, they meant reducing emissions. Since that time, vulnerable nations and frontline communities have pushed adaptation onto the global agenda. We are already being affected by climate change, they argue, so we need to not just prevent but also treat the current harm being done.

More recently, the most affected countries and populations have pushed for compensation. Not only have we failed to prevent climate change, they argue, but its impacts are already so great that they cannot be adapted to. Low-lying islands, for whom even a small degree of climate change is existential, have been strong advocates for so-called loss-and-damage measures in international climate policy, demanding that those who have contributed most to climate change pay the reparations. In the future, expect these claims to grow.

And as climate change proceeds, what was previously unthinkable may become widely demanded. Today, many climate advocates reject geoengineering techniques as an unproven distraction from mitigation efforts. But if the impacts of climate change continue to accumulate, governments may come to see such technologies as vital components of national security.

All of these strategies will be far more costly and far less effective than mitigation. But by the time today's school strikers are watching their own children take to the streets, these strategies might be seen as the only options left.

For this reason, a world increasingly shaped by climate change will likely have strong incentives for *more* short-termism, not less. This paradox is another of the various cruel ironies of climate change: even as it shows us the need for longer term governance, it undercuts our ability to reform in that direction. This makes climate change an ever more dangerous threat than is widely recognized since it attacks precisely the political support for longer term governance functions that can best address it.

Can we, as a society, get back in the driver's seat? Historically, political institutions tend to lag social and material conditions. On the macro level, thinkers like Marx or Polanyi saw the development of modern governance as a reaction to the rise of industrialization and capitalism. Marx, looking at the start of the process, got the outcome wrong; Polanyi, with the advantage of hindsight, got it right. But both fundamentally see changes in political institutions following changes in the economy and society. Contemporary scholars have posited that that emergence of systemic environmental challenges may lead to similar transformations in political institutions.[10] "Governance in the Anthropocene" may indeed be quite different from what we have now, but if existential climate politics force us toward greater short-termism, there is no guarantee that the governance we need is the governance we will get.

Instead, the challenge of climate change, and of long problems more generally, is not to wait for human governance to be shaped by our new realities but to get at least somewhat ahead of them. This is a radical idea but perhaps not an impossible one.[11] The strategies laid out in this book—being advanced all around the world—offer a path not of reaction but of anticipation. It lies within our collective choice to adopt them or not. In this sense, climate change is, if not a driver that will inevitably lead us to better long-term governance, a potential catalyst that we can use to move in that direction.

Indeed, the arc of our struggle with climate change is arguably at present in a kind of optimal zone for substantial reform of political institutions (compare the scenarios discussed in chapter 6, especially figures 6.3 and 6.4). Although we have missed the best scenarios, there is still an enormous amount we can do to prevent further climate destabilization, and we still have time to adapt and build significant resilience to the impacts already coming our way. At the same time, the devastating impacts of climate change are now widely felt, creating more urgency and salience than has previously existed. Together, these two conditions create a unique window of opportunity for early action. Put another way, the interests of decision-makers and decision-takers overlap more than they have in the past or will in future. Decision-makers can now seize this moment to put in place the institutional changes that will tilt the balance of power toward pro-long-term interests for decades and beyond. But if they fail to act in advance, the immediate, urgent impacts of climate change will progressively sap their ability to do so, tilting politics further against effective governance of long problems. Now is the time to act.

Learning to Fly

Imagine we succeed. Suppose we do not allow climate change to drive us to more reactive politics, but instead we use the moment of opportunity climate change creates to catalyze longer term governance across all areas of policymaking. In this scenario, governments create a range of long-term-oriented institutions that help them address climate change and in doing so, also ameliorate the challenges that the early action paradox, shadow interests, and institutional lag pose to all long problems. Climate-focused long-term governance tools become more general and embedded across governments, covering all policy areas. Over time, building blocks like the UN Declaration on Future Generations catalyze profound changes in governance equivalent to the rise of democracy or multilateralism. Climate change could serve as a catalyst for the kind of future-oriented government that thinkers and activists have called for.[12]

In this scenario, going back to Vickers's metaphor, human societies are not just hurling themselves into the great unknown, caring not where they may land. Instead, we look ahead and, at least to some degree, control and navigate our path through history. Instead of free-falling, we clumsily, haltingly, imperfectly learn to fly.

I do not assume this is the most likely scenario. Some of the reforms this book proposes seem difficult to imagine from where we sit now. But the long-term perspective helps us see that whether something is immediately feasibility or not is not always the right question to ask. Instead, we should ask if we can change the direction of travel in our political institutions toward more long-termism or if we can accelerate the pace of change in that direction. This book shows that we can. Indeed, many of the first steps we need to take to move in that direction have already happened.

Still, is it pure hubris to imagine human societies gaining more control over our collective future? History urges caution. During the modern period, advances in science, technology, and human organization led many to embrace an idea of rational progress as a core project for humanity. By the middle of the twentieth century, the unprecedented violence of two world wars, the invention and deployment of civilization-ending weapons, and the recognition of the precarity of planetary systems, among many other developments taken in the name of progress, led many to question the idea that rational steps toward a better future were possible or even desirable.[13] But with long problems like climate change now demanding unprecedented shifts in the economy over many decades, the imperative to regain some control over the "free fall" of human development compels us to at least try.

If this regaining of control is possible, it will only happen through governance. Vickers is clear: "Our major instrument of adaptation is government."[14] Science fiction writers typically imagine fantastical technologies like time machines or eternal youth elixirs that will give humanity new powers, for good or ill. But taking long problems seriously shows that if human societies are really to master time, they will need better governance systems, better "political technologies." To the extent we succeed, long-term governance may not allow us literally to travel across time or live forever. But it will create the potential for human civilization to collectively embark on projects of greater and greater temporal scale. In that possibility lies not only the dangers and challenges this book has grappled with but also the hope of what human endeavor may yet achieve.

APPENDIX 1

Why We Face More Long Problems Now

The temporal frame of political problems is changing in two seemingly contradictory but in fact related ways. On the one hand, we face more problems in which causes and effects spread over significant time spans. On the other hand, the rate of many social and political processes is accelerating, increasing the short-term pressures that shape human activity and especially political action. This combination of lengthening problems and accelerating social processes creates a mismatch between the timescale on which we govern and the timescales on which the objects we seek to govern unfold.

This appendix and the next sketch these two intertwined trends, drawing on the literature in earth sciences, ecology, sociology, philosophy, and normative political theory that have dealt with them. Although a full treatment is beyond the scope of this book, my hope is to show how these trends require greater attention from social scientists and policymakers.

Why might we have more long problems than in the past? I outline three drivers—really, two drivers and one emerging factor—that make problems longer than they were in the past, following the elements of political problems defined in chapter 1 and repeated in table A.1. First, material realities have shifted as the scale of human development increasingly brings human systems into contact with planetary systems—like the carbon cycle—that operate on radically different timescales. Second, technological and scientific development have further altered both our material ability to affect the distant future as well as our ability to measure and understand problems beyond the present. And third and only tentatively, social beliefs about how to value future generations have begun to shift.

Table A.1. Why we face more long problems now

Driver	Problem elements it relates to	Mechanisms
Collision of human and planetary systems	Material or social facts	Human development alters planetary systems, creating existential risks for humanity
Rapid change in technology	Technical and scientific processes Material or social facts	We can perceive and understand cause and effect over the long term Heightened ability to substantially affect future generations
Moral pantemporalism (partial)	Social understandings and preferences	We (should) value well-being of future generations

To be clear, not all problems are getting longer. Some are staying the same, or even getting shorter (the next appendix turns to acceleration). The argument is rather that we face more long problems today than we did in the past because some problems are getting longer and becoming more politically salient.

The Collision of Human and Planetary Timescales

The year 1970 marked an inflection point in human politics. In that year, the celebration of the first Earth Day indicated the emergence of the modern environmental movement and, soon thereafter, modern environmental regulation and global environmental governance. But 1970 was also the first year humanity used up its entire annual "budget" of natural resources in less than 365 days, according to Global Footprint Network.[1] This organization seeks to calculate the total ecological footprint of our civilization each year and declares Earth Overshoot Day to be the date on which that footprint exceeds the earth's natural carrying capacity. The first Overshoot Day fell on December 29, 1970, meaning humanity was using just a few days' more resources than the planet could supply. By 2020, the date came forward to August 22. At that level of usage, we need 1.75 earths to sustain ourselves. In other words, 1970 or thereabouts marked the beginning of a new phase in human history during which the boundaries of the planet itself have become relevant matters of political concern. While the Overshoot Day framework is of course only one way to assess and represent this idea, it puts a novel and complex concept into clear terms: the material facts around our environment have shifted.

The reason for this stark new material reality is the fantastic pace and scale of human development in the postwar period. The economic rebound that followed World War II began an extraordinary expansion of the global economy as, over decades, industrial production spread around the globe and living standards rose rapidly. Adjusted for inflation, global GDP is about ten times greater today than it was in 1950, the population has more than doubled, and the average person is over four times richer in real terms.[2] The urban population and primary energy use have both grown fivefold.[3] Human development sped faster than anything we have experienced in our one hundred thousand years as a species.

This "Great Acceleration" has fundamentally changed the material facts around humanity's role in planetary ecosystems. The International Geosphere-Biosphere Programme coined this phrase to describe the growing imprint of human activities. From 1950 to 2010, usage of freshwater and loss of tropical forests more than doubled, the number of marine fish caught increased fivefold, the number of large dams increased six times, and fertilizer consumption grew over one hundred times.[4] Global emissions of carbon dioxide have grown sixfold since 1950, changing the composition of the atmosphere.[5]

The impacts on planetary systems have been equally monumental. Biodiversity has plummeted, with one leading index showing a 68 percent reduction in species prevalence since 1970.[6] We have now caused the most recent of several mass extinction event in Earth's history.[7] Other essential natural processes like the phosphorous cycle or ocean chemistry have been disrupted by our agricultural expansion and carbon emissions, respectively. One influential framework identifies nine "planetary boundaries" across different earth systems and finds that we have already breached at least four of them.[8] Perhaps most ominously, the average temperature of the planet has already warmed more than 1°C above preindustrial levels and is on course to warm more than twice as much again by the end of this century. The direct effect of this change on sea levels, weather patterns, heat, water scarcity, biodiversity, and food production will fundamentally alter nearly every economic system. And their knock-on effects on, for example, migration, disease, and war promise to create a threat of existential proportions.

Although previous civilizations have faced localized ecological crises—think of Rapa Nui (Easter Island)—for the first time in human history, we are affecting the fundamental systems of the planet as a whole.[9] As a result, geologists have proposed declaring an end to the Holocene, which began after the last Ice Age, and announcing the start of the Anthropocene, an era in which the actions of humanity are the dominant drivers of planetary outcomes. While debates on exactly how to define the Anthropocene and calculate its start point (in the twentieth century, with the Agricultural Revolution?) remain ongoing, there is little doubt that human and planetary systems have collided in a startling new way.

Given the scale of these disruptions, it would be very surprising if our politics remained unchanged. But social scientists have only recently really begun to grapple with the implications of this collision for politics and governance.[10] Prominently, students of global environmental politics have proposed a new framework of "earth system governance" that would include ecosystem- and global-level decision-making bodies, armed with science, to match this new reality.[11] While the spatial and scientific dimensions have received perhaps the most attention in this growing literature, the "temporality" of governing in the Anthropocene has also emerged as a key area of interest.[12]

To see why, the philosopher Helga Nowotny's concept of "proper time" can be helpful.[13] Nowotny argues that different social systems operate via their own time frames, and so each part of society has a "proper" time in the sense that it experiences time in a way that is appropriate and unique to itself, even if we all now use the same units of time. Analogous to how objects moving at different speeds will, according to relativity, experience time differently, different social contexts have their own "senses" of time. In Nowotny's view, work time is different from family time, and time as experienced by preindustrial societies is very different from our current conceptualization.[14]

Extrapolating beyond human systems, we can therefore expect the collision of human and natural systems to create an even more striking temporal mismatch.[15] The geologist Marcia Bjornerud laments that even well-educated people today are ignorant of the Earth's long history and so struggle to understand the "intrinsic timescales" of the natural world, much less govern in a way that takes them into account.[16] Industrial society is only about two hundred years old, but its impacts have already shaped the planet for millennia to come. The species we have driven to extinction will take millennia to be replaced by new forms of life. Various inorganic substances we have put into the natural environment, like plastics, will continue to appear in the stomachs of fish and birds—and those of our descendants who eat them—for thousands of years. And, of course, carbon emitted by our ancestors generations ago will continue to warm the climate in which our descendants many generations hence will live.

In sum, the material facts have changed, the first element of the problem framework developed in chapter 1. The causes and effects of natural systems are temporally distant, spanning time periods that we find difficult to fathom but now are forced to grapple with. The material facts we confront have gotten longer as human-induced stresses on the natural environment now force us to govern around processes that operate on a "proper time" radically longer than our own.

Science and Technology Allow Us to See the Future More Clearly and to Affect It More Decisively

Shifts in material facts are not the only things that make problems longer, of course. The second element of political problems—our scientific and technological understandings of material and social facts—can also expand the timescale on which we perceive proximate causes and effects and therefore how we define problems. Consider the examples of climate change and a forest fire given in chapter 1. To someone without knowledge of climate science, a forest fire can only be seen as a forest fire, a near-instantaneous problem. But once we know that forest fire frequency and intensity are affected by the concentration of carbon emissions in the atmosphere, the timescales on which we perceive the problem shift. In a similar vein, Bjornerud notes, "Early in an introductory geology course, one begins to understand that rocks are not nouns but verbs— visible evidence of processes: a volcanic eruption, the accretion of a coral reef, the growth of a mountain belt."[17] Even if the material facts of climate change or forest fires—or rocks—are the same regardless of whether we understand them or not, our scientific and technical ability to theorize and measure extends our understanding of the temporal span of cause and effect.

For climate change, this ability to perceive the problem should not be taken for granted. Comprehending climate change has required an accumulation of scientific and technical information by thousands of people over decades, what Paul Edwards terms a "vast machine" of theory, models, and observational data from such diverse places as millennia-old ice cores, dusty, centuries-old weather notes in dozens of languages, and satellites armed with cutting-edge sensors that take measurements multiple times a second.[18] Historians of technology refer to such knowledge systems as an accumulative infrastructure. Modern meteorology depends on concepts we now take for granted but that represent fundamental breaks from the past, such as global cartography, a universal system of timekeeping, and thousands of real-time, networked observation points. Before the advent of modern technology, we could perhaps understand changes in the weather on longish (at most, several generations) timescales.

But only by fitting all this information together into a global, transhistorical model of the climate can we begin to perceive of climate change as a problem to be solved, as opposed to just "the way things are."

Edwards's "vast machine" has a political dimension as well. As scientific understanding of global warming grew in the 1980s, governments and international organizations saw a need to aggregate and evaluate the global state of knowledge on this complex phenomenon. In 1988, the World Meteorological Society and the United Nations Environment Programme created the IPCC, tasked to assess and summarize the state of knowledge on climate change. The IPCC's regular reports are massive undertakings. The fifth assessment report, released in three parts over 2013 and 2014, involved over 830 lead authors and reviewers summarizing the work of one thousand contributors, with the output then being reviewed by a further two thousand experts.[19] But this work is not only technocratic. Governments have reversed for themselves the right to review and vet the IPCC's publications before release, particularly its influential summaries for policymakers. These are fundamentally political contests over problem definition.[20] What aspects of the problem the IPCC chooses to emphasize and highlight powerfully structure how we understand the problem and its potential solutions.

Climate is hardly the only area where greater powers of observation and understanding have changed the way we understand problems.[21] We have built similar "vast machines" of data collection, theorizing, and modeling to understand global health challenges, such as the surveillance system for pandemics. An even more complex technical ecosystem monitors and measures the global economy. Note, however, that this increased observation and understanding does not always lengthen problems; it may also shorten them. For example, the ability to perceive economic indicators at scale in real time has contributed to the development of faster and faster ways for market traders to arbitrage, shortening time horizons to generate both profits and risks. In a pandemic, real-time data are essential for understanding the spread of a disease and measures to counter it. But in climate, as in many areas, expanding our understanding also reveals new drivers, many of which unfold on transhistorical timescales. For example, advances in bioinformatics have allowed us to sequence the genomes of harmful microorganisms, which, combined with global datasets, has helped us to understand the development of antimicrobial resistance over time as a long problem. Bacteria are evolving to evade our most sophisticated antibiotics. Whether deeper understandings of a problem lengthen or shorten that problem ultimately depends on the other elements that comprise its structure and especially the underlying material and social facts.

That caveat aside, students of the history of science and technology have noted other ways in which increasing interest in, and capacity for, seeing across time has lengthened the way political problems are defined. Scholars identify an inflection point in the middle of the twentieth century, during the postwar period of high modernism in the two decades after World War II.[22] Several trends made this a particularly febrile period for "futurism" of various kinds. First, early computers began to make data analysis feasible at new scales, a key part of the "vast machines" described above. Second, the advent of the Cold War and the nuclear age created an overwhelming imperative to understand when war might break out and how it could be won. Scientists at RAND and other institutions began to develop complex models to look into the future. The modern field of statistics advanced quickly, and probabilistic methods began to spread to a wide array of academic disciplines, meaning "the future, which had been discussed as a more and philosophical category since the seventeenth century, became an object of social science."[23] Third, this period marked a high point of government interest in planning, in both the capitalist West and communist East and, through new theories of economic development like modernization theory, vis-à-vis so-called third world countries. Finally, the rise of modern environmental problems, discussed above, expanded our view of the timescales of the objects we sought to govern.[24] The influential Club of Rome report *The Limits to Growth* presented a first model of multiple natural systems at planetary scale, introducing ideas of planetary boundaries.

In these various ways, changes in science and technology have contributed to the lengthening of political problems by changing our understanding of cause and effect. To be sure, they have also had the effect of accelerating various processes, as I discuss in the following appendix.

But first, it is important to return to the first element of political problems—material and social facts—which have also been affected by the broader changes in science and technology. The core argument is that we do not just see the future more clearly; we can also affect it more significantly.

A significant strand of sociological literature, including many popular treatments, has explored the idea that the increasing rate of technological change makes the future less and less like the past.[25] It is often observed that a peasant living in Europe in the fourth century could expect her descendants in the fifth century to have essentially a similar range of life experiences. Wars, famines, and plagues may come and go, but the way people lived and worked, their beliefs, and the kinds of political systems that governed them could be expected to remain the same.

The modern age is quite different. Technological change, entwined with social change, makes experiences vary significantly from one generation to the next and even within a life span. The faster the rate of change, the less certainty we have about what the future might look like but also the more agency we have to shape that future. This is why social theorists like Nowotny believe that we have "destroyed the future," as it was previously understood, and instead collapsed it into an "extended present."[26]

We can see this idea clearly in the case of environmental change. The choices we make about the climate today will create fundamentally different states of the world for future generations. Students of emerging technologies like gene editing or AI have made similar claims, as indeed, an early observer of Gutenberg's printing press or a contemporary observer of social media might have. Simply put, accelerated rates of change make our choices much more consequential for the future generations than those of Dark Age peasants were. To be sure, if they made the wrong choices—for example, neglecting to prepare for a bad harvest or forgetting to reinforce the fortifications against raiders—they could condemn their village or kingdom to suffering or death. But if we make the wrong choices—for example, with nuclear weapons, AI, or climate change—we now have the potential to affect humanity and the planet as a whole.[27]

This change has important political implications, perhaps best articulated by Ulrich Beck's influential idea of a "risk society."[28] New technologies generate benefits but also create new harms and risks, many perhaps unintended. For example, in the 1950s, many believed nuclear power would provide limitless energy, while underestimating or ignoring difficulties around safety and disposal. More recently, advocates of social media anticipated a revolution in human communication that could underpin a deeper, digital democracy, without foreseeing the potential for echo chambers and strategic misinformation to instead undermine traditional forms of democratic politics or enable authoritarian control. The risks around new technologies are often long problems in that their effects only become apparent long after they become widespread.

Governing such risks is enormously difficult, Beck writes:

> It is true: the dangers grow, but they are not politically reforged into a preventative risk management policy. What is more, it is unclear what sort of politics or political institutions would even be capable of that. . . . The openness of the question as to how the dangers are to be handled politically stands in stark contrast to the growing need for action and policymaking.[29]

Because traditional political institutions struggle to adequately manage these changes, Beck argues, political decisions shift, de facto, from the organs of the state to scientists, technocrats, and the private sector, where decisions regarding technological innovation are made.[30] It is these actors and institutions who make the choices that shape the future in which subsequent generations must live. Politics and governance must therefore find new ways to ensure decisions made in these domains reflect the long-term public interest. For Beck, technological changes create a new class of long problems—managing their unforeseen risks—that our society is poorly equipped to handle. This is a change in the material conditions we face. As Boston puts it, "The growing capacity of humanity, through rapid technological advances, to cause not only more widespread and severe harm but also harm that is persistent, if not irreversible," is a largely unprecedented kind of problem.[31]

In sum, scientific and technical advances make problems longer in two ways. On the level of understanding, they enhance our ability to see longer term processes like climate change. On the level of material facts, they increase the weight of decisions made today on the outcomes—intended or otherwise—that society will experience tomorrow. Ironically, the same abilities that increase our powers of prediction also make the world less predictable in that they increase the influence of a highly variable factor: human agency. But greater power to affect the future can also, perhaps, be positive. As a 2013 report by the UN secretary-general put it, "Scientific inquiry allows society to understand the long-term impacts of its actions, while technological advancement means that it is in a position to mitigate harmful consequences if it so chooses."[32] The *Whole Earth Catalog* tagline put it more succinctly: "We are as gods and might as well get good at it."

A Changing Normative Frame?
The Growing Moral Shadow of the Future

The third element of political problems—actors' preferences and beliefs—can also lengthen problems. If we "care" more about the future—either for moral or material reasons—problems with more temporally distant effects rise in importance.

Social scientists typically model the extent actors care about the future as a discount rate, which reduces the value of a particular thing or outcome over time. I consider this concept in chapter 6. Game theorists refer to the expected "payoffs" from future outcomes as the "shadow of future," which, when significant—that is, when the discount rate is low—can change behavior in the present. For example, if you know you will negotiate with a partner again and again in the future, you may have less incentive to cheat them in the present, as that might harm future deals—a dynamic game theorists term the folk theorem.

Although conceptually clear, how much do we actually discount the future in practice? Measuring empirically how much humans value the future is challenging, and we cannot say with certainty how this has changed over time.[33] Given these complexities and uncertainties, there is no systematic evidence that we now care more about the future than in previous periods. Indeed, the literature highlights numerous examples of rising short-termism, especially in the economic sphere, discussed in chapter 2. We therefore cannot say that we see evidence of discount rates falling; they might even be rising.

But leaving aside the empirics, philosophers and political theorists are increasingly making a strong case for why we *should* care about the future. Observing the rise of long problems, these voices argue that we should, as the philosopher Roman Krznaric pithily puts it, strive to be "good ancestors."[34] To the extent they are right, long problems hold greater moral weight.

This book does not fully assess the normative argument about the extent to which we should care about the well-being of future people but rather focuses on the implications of the idea for politics and governance. A helpful analogy can perhaps be drawn to shifting ideas about how we should care about people across space, which shares a common core with many of the normative arguments advanced for future generations. Since at least the Enlightenment, moral philosophy and political theory have increasingly emphasized the intrinsic value and equality of each individual human being, regardless of where they are born. As the world grew increasingly globalized, philosophers argued our ethical obligations to distant people and places must also increase. After all, in an interdependent world, our actions affect people far outside our city, region, or nation, and their actions affect us. Should we not consider their well-being (and they ours), and should politics and governance not be structured so as to ensure we all do? As David Held and other advocates of such moral cosmopolitanism have written, in a globalized world, our "shared communities of fate" now span the globe, creating a spatial disconnect between our increasingly transnational moral sphere of concern and our traditional, jurisdictionally divided political institutions.[35] Climate change—in which each of our individual actions affects the climate that all others must live in—demonstrates this idea particularly clearly.[36]

Moral cosmopolitanism seems utopian, but that does not mean it has not had a powerful political influence. Indeed, political institutions have shifted along with these ideas in ways that would likely have been unimaginable at most stages of human history. Significant political documents have declared universal human values since at least the eighteenth century, such as the 1776 American Declaration of Independence or the French Declaration of the Rights of Man and of the Citizen in 1789. In the twentieth century, these ideas advanced significantly with the creation of a modern international human rights law and institutions. The fundamental equality and inalienable rights of human beings, no matter where they are born, is universally recognized in dozens of international treaties, national constitutions, and other institutions. Some principles have even begun to challenge traditional understandings of national sovereignty, such as the idea of universal jurisdiction (all national courts have a right and responsibility to try crimes against humanity) or the responsibility to protect (a state's sovereignty is conditional on its ability and willingness to protect the fundamental rights of its citizens).

As these examples highlight all too clearly, shifts in moral ideas *can* affect political outcomes and governance structures, but there is no guarantee they *will*. Moral cosmopolitanism is a powerful idea that has informed real changes in politics and governance. But the world that exists today is far from approximating even a minimum realization of its ideals. We may see problems as "wider" than we would have before, in part because we recognize people who live far away as holders of rights, but that does not necessarily imply we are better at solving them.

The "time revolution" that a range of philosophers now call for is likely similar.[37] As Toby Ord argues, "Just as it would be wrong to think that other people matter less the further they are from you in space, so it is to think they matter less the further away from you they are in time."[38] The implications of this longtermism—the idea that each human life is intrinsically valuable, no matter when someone is born—are perhaps even more radical than those of cosmopolitanism. Ord notes that if *Homo sapiens* last as long as the typical mammalian species (about one million years), the vast majority of humans are yet to exist. Acting in a way that treats them as our moral equals, protecting their fundamental interests, is a stark challenge to those of us in the present. To the extent we accept it, the problems we seek to govern will be very long indeed.

In practice, these ideas seem perhaps even further from our reality than the idea of giving everyone currently on the planet an equal say in our collective decision-making. But it is important to note that we can find references to the importance and value of future generations in a welter of political documents today. The 1987 UN report *Our Common Future*, which brought the term "sustainable development" into the policy mainstream, defined that concept as meeting the needs of the current generation without compromising the ability of future generations to meet their needs. A recent study found that over 40 percent of existing constitutions refer to future generations, a figure that has grown steadily over time.[39] A 2013 report from the UN secretary-general noted that in sum, such statements indicate a collective emphasis on the moral value of future generations.[40]

Moral long-termism is therefore the last and most partial reason why long problems are more salient today. To the extent such ideas spread and take root, we can expect the moral shadow of the future to lengthen our political problems, just as cosmopolitanism has widened our conception of problems. As that example shows, we can expect this process to be partial and halting but still important. We should not discount the potential of such norms to spread and harden. Indeed, over the long term, these changes in fundamental social beliefs may become one of the most important factors in determining the temporal span of political problems.

APPENDIX 2

The Role of Acceleration

For the reasons discussed in appendix 1, political problems grow longer. But many have argued they are also getting faster, which at first blush, seems contradictory. It is therefore useful to address the so-called acceleration hypothesis, the idea that social and political processes unfold more rapidly than in the past. Acceleration can mean that some problems indeed get shorter, to the extent the temporal gap between cause and effect shrinks. But I argue that acceleration and lengthening are not necessarily contradictory and can even have a complementary, and pernicious, impact on politics and governance. To the extent acceleration causes political processes to shorten, they become even more mismatched to the needs of long problems.

Again, it is sociologists and social theorists who have most developed these ideas. In Hartmut Rosa's seminal treatment, acceleration is fundamentally linked to modernity and comprises three processes: technical changes like faster communications and transport, the increasing rate of social change, and what he terms the quickening pace of life, the sense that we never have enough time.[1]

It is easy to see how these forms of acceleration can shorten political problems by shrinking the temporal distance between cause and effect. In the Middle Ages, it could take months for a deadly pathogen to spread across footpaths and sea-lanes. Today, a global pandemic can materialize in weeks. Military conflicts used to take months or even years to begin. Countries would need to muster their forces across large territories and maneuver them to the field. Today, military strikes can be ordered and executed in minutes or, with cyber warfare, essentially instantaneously. The advent of social media means political scandals can emerge, metastasize, and disappear all in the space of a single "news cycle." The speed of many political problems means that governance has to deal with more problems, more quickly, than in the past.[2]

Although "short problems" are interesting in and of themselves, the relevant element for this book is the interaction between acceleration and long problems. Most obviously, immediate problems tend to shift attention and effort away from long problems. But speeding things up can have further destabilizing effects. A core concern of what we might call the acceleration literature is that a quickening rate of change leads to societal disorientation. Alvin Toffler's 1970 bestseller *Future Shock* popularized the idea that rapid change— "the process by which the future invades our lives"—exceeds the capacity of political and social systems, and even individual psychology, to cope.[3] Toffler's investigation of this "sickness" of modernity built on William Fielding Ogburn's 1923 theory of cultural lag. For Ogburn, the rate of technological change is roughly a function of the accumulated stock of technology and therefore increases exponentially. Culture (he uses the term broadly), on the other hand, shifts only slowly. For example, he offers a detailed study of how nineteenth-century industrialization led to more and more workers' deaths in accidents, while safety laws and practices lagged decades behind.[4] In the political realm, these ideas are reflected in Samuel Huntington's famous critique of modernization, his worry that too rapid economic modernization would exceed the rate at which new interests could be incorporated into politics and therefore lead to political disorder.[5] But even earlier observers of the modern era noted similar trends. One of the most quoted passages of the *Communist Manifesto* observes that with industrialization, "all fixed, fast-frozen relationships, with their train of venerable ideas and opinions, are swept away, all new-formed ones become obsolete before they can ossify. All that is solid melts into air."[6]

This destabilizing acceleration holds many potentially negative consequences for governance. In what Paul Fawcett terms the accelerated polity, "fast" policymaking challenges traditional Weberian bureaucracies, based on formal processes, and more critically inhibits deliberation, due process, and the time-consuming work of collective decision-making, which are better served by "slow" policymaking.[7] In this sense, acceleration poses a particular challenge to democracy, many argue.[8] It also poses a decisive challenge for governing long problems, which require long-term planning and often short-term sacrifices for long-term gains. For example, planning complex projects like the decarbonization of a power grid or the restoration of flood-absorbing natural marsh requires consistent, additive action over decades and so is poorly served by rushed decisions.

If acceleration operates in parallel to lengthening, it is important to understand how they combine. Although the two trends push in opposite directions—some problems lengthen, others shorten—they also have areas of overlap and interaction. Both are linked to the process of rapid technological change. Together, they push governments toward shorter and shorter time frames even as policymakers need to be looking further and further ahead.

Writing in 1970, Geoffrey Vickers summed up these competing demands well:

> The content of our political system—the sum of relations which we aspire to regulate—has grown and is growing in volume; and the standards to be attained have risen and are rising. The action needed to attain and hold these standards requires more massive operations, supported by greater consensus over far longer periods of time than in the past. On the other hand, the situations which demand regulation arise and change with ever shorter warning and become ever less predictable, as the rate of change accelerates and the interaction variables multiply. Clearly the task of the political regulator becomes ever more exacting.[9]

In other words, the combination of acceleration and long problems is not oxymoronic but rather linked and dangerous.[10] Again, climate change aptly demonstrates these pressures. Long-term planning and vision is required to understand the problem, develop solutions, and implement them at scale. At the same time, because every day of delay makes the problem more difficult to solve, immediate action grows ever more urgent. As Klaus Goetz puts it, referring to financial crises, politics must somehow be both responsive and responsible.[11] We need "fast, long" politics but instead, too often find ourselves trapped with "slow, short" politics.

In sum, to the extent we face more long problems today, the dangers of acceleration become greater. Writing in 1923, Ogburn fretted,

> Never before in the history of mankind have so many and so frequent changes occurred. These changes, it should be observed, are in the cultural conditions. The climate is changing no more rapidly, and the geological processes affecting land and water distribution and altitude are going on with usual slowness.[12]

If only that were true. Since Ogburn wrote those words a hundred years ago, we have changed the climate by more than it has changed naturally in the previous ten thousand years.

How to respond? The "acceleration hypothesis" logic of Rosa, Toffler, and Ogburn is essentially Malthusian. Just as Malthus thought agriculture could support only so many humans, students of acceleration fear that human systems can change only so much so fast. But just as technological innovation ultimately proved Malthus wrong (see chapter 6), is it possible that social and political innovation—better governance—might allow humanity to adapt to temporal stresses we have brought on ourselves? For Rosa, Toffler, and Ogburn, that is the pressing question. Rosa sees a need to shift democratic forms of control from traditional spheres of policymaking to businesses, social movements, or online communities.[13] Ogburn sees government as the critical tool of adaptation. Toffler argues for nothing less than new forms of "anticipatory democracy." Governing over time requires innovations. This book attempts to outline some ways forward.

NOTES

Chapter 1: Long Problems

1. Orr 2016.
2. Boston 2016.
3. Held 1995.
4. Ostrom 2009.
5. Lowi 1964.
6. Kendi 2017.
7. Jacobs 2011, p. 5.
8. IPCC 2014.
9. Oreskes & Conway 2010. The aim is typically not to "disprove" theories of global warming but rather to raise enough questions about its causes, speed, and impact to neuter the onus for prompt action. Because science is a system of "organized skepticism" grounded in challenge and dispute and because the climate system is immensely complicated and difficult to study, this strategic use of misinformation and doubt can be incredibly effective, at least in the short- to medium term, for delaying action. For example, my colleagues and I analyzed the text of all the quarterly earnings calls of major oil and gas companies from 2005 to 2019. We found that the majority of managers' statements regarding basic climate science expressed skepticism or doubt as late as 2014, reflecting how, for decades, the material facts of the problem were themselves contested. Green et al. 2022.
10. Barrett 2003; Victor 2011.
11. Geels et al. 2017.
12. Aklin & Mildenberger 2020.
13. Bernstein & Hoffmann 2019.
14. Gates 2021.
15. Colgan et al. 2020.
16. Schipper 2006.
17. Mitchell 2006.
18. Oye 1986; Koremenos et al. 2001.
19. See Pierson 2004.
20. See Pierson 2004, p. 80.
21. Kelley et al. 2015.
22. MacKenzie 2021b.
23. MacKenzie 2021a.

24. Pierson 2004, chap. 3.
25. MacKenzie 2016.
26. Following Ostrom, Levin et al. discuss climate change as a "super wicked" problem defined by four characteristics: time is running out, those who cause the problem are also those who must address it, central authority is weak, and policy responses discount the future irrationally. Boston sees climate as representative of a class of "creeping problems," which he defines as having a slow and incremental nature, a lag between cause and effect, and various path dependencies. Levin et al. 2012; Boston 2016.
27. Vickers 1970, p. 69.
28. Vickers 1970, p. 70.
29. As one of the leading thinkers on governance and time, Sheila Jasanoff writes, drawing on German scholar Ulrich Beck in the modern era, "What societies seek to govern, and how they govern it, were transformed. The aims and objects of governance morphed in size and scope, as well as in their temporal situation, from small and near-term problems—those potholes and snowy roads—to distant, geographically dispersed, and increasingly faraway futures." Jasanoff 2020, p. 34.
30. Steffen et al. 2004.
31. Bjornerud 2018.
32. Edwards 2010.
33. Macaskill 2022; Ord 2020; Krznaric 2020; Kim & Harrison 1999.
34. Macaskill 2022, p. 9.
35. Araújo & Koessler 2021; Krznaric 2020.
36. Martínez & Winter 2023.
37. Wenger et al. 2020.
38. Beck 1992; Andersson 2018; Jasanoff & Kim 2015; Jasanoff 1990.
39. Porter & Stockdale 2016.
40. Stockdale 2016.
41. Hom 2018.
42. Bernstein & Hoffmann 2019; Levin et al. 2012. For a noncritical approach, see also Drezner 2021.
43. Biermann et al. 2012; Delanty & Mota 2017; Milkoreit 2017; Biermann 2014; Bornemann & Strassheim 2019.
44. For example, Simon Caney (2022b) provides a cogent assessment of future-oriented institutions in global climate governance.

Chapter 2: Why Long Problems Are Hard to Govern

1. Boston 2016.
2. Bjornerud 2018, p. 7.
3. Lavery & Donovan 2005.
4. For a summary, see Boston 2016, p. 67.
5. For a summary, see Boston 2016, p. 72.
6. Boston 2016, p. 5.
7. MacKenzie 2021a; Smith 2021.

8. Libman 2020.
9. Libman 2020.
10. Krznaric 2020, p. 173.
11. Finnegan 2022.
12. Simmons 2016.
13. Zhou & Wang 2020. For a general discussion, see Wu et al. 2017.
14. Jones 1994.
15. Allison & Halperin 1972.
16. Jacobs 2011.
17. Jacobs & Matthews 2012.
18. Bernstein & Hoffmann 2019.
19. Boston 2016, p. 60.
20. Krznaric 2020.
21. Johnson 2018.
22. MacKenzie 2016, 2021a.
23. United Nations Secretary-General 2013.
24. Helg 2019.
25. Drescher & Emmer 2010.
26. Keck & Sikkink 1998.
27. Held 1995.
28. Coase 1960.
29. This stylized framing of mitigation present and future costs and benefits has been critiqued in the literature (see the transition-focused theories described in chapter 6) but provides a helpful analytic foil.
30. Goldstein & Gulotty 2021.
31. Boston 2016; Jacobs 2016.
32. Krznaric 2020, p. 7.
33. Fraser 1999, p. 30.
34. Keohane & Victor 2011.
35. Abbott 2012.
36. Grantham Research Institute on Climate Change and the Environment 2021.
37. UNFCCC 2019.
38. Hale et al. 2013; Hale & Held 2011; Abbott et al. 2016.
39. Koremenos et al. 2001; Koremenos 2016.
40. Jupille et al. 2013.
41. Jupille et al. 2013.
42. Hale et al. 2013.
43. Pierson 2000.
44. Pierson 2000.
45. Greif 2006.
46. Ikenberry 2000.
47. Pierson, 2000, p. 114.
48. Note that institutional lag can also bedevil what chapter 2 defined as ongoing problems—those that reoccur over time but have shorter spans between cause and effect.

49. Koremenos et al. 2001.
50. Tarsney 2022.
51. Rosa 2013, p. 264.
52. Lindblom 1959.
53. Johnson 2018.
54. Valenzuela 2020.
55. Stokes 2020.
56. Collingridge 1980, p. 11.
57. Flood 2013.
58. Pierson 2000; Wegrich 2019.
59. Hill & Martinez-Diaz 2019.
60. Royal Dutch Shell 2021.
61. Galik 2020; Smith et al. 2023.
62. Hale et al. 2020.

Chapter 3: Forward Action

1. Lindblom 1959, p. 87.
2. Jasanoff 1990; Wenger et al. 2020.
3. Guston 2014.
4. Fuerth & Faber 2011, p. 1.
5. Wegrich 2019.
6. Venkataraman 2020.
7. Edwards 2010.
8. Culpepper 2010.
9. Helleiner et al. 2010.
10. Sachs et al. 2022.
11. Howe et al. 2019.
12. Gazmararian & Milner 2022.
13. Howe et al. 2019.
14. Mbaye et al. 2021.
15. Skocpol 1996.
16. Section 101(b).
17. Wood 2002.
18. Boston 2016, p. 388.
19. Jacob et al. 2011.
20. Nordhaus 2017.
21. Boston 2016, chap. 12; Wenger et al. 2020; Chermack et al. 2001; Amer et al. 2013.
22. Horowitz 2020.
23. Tetlock 2005, p. xxi.
24. Horowitz 2020.
25. Dreyer & Stang 2013; Prítyi et al. 2021; Greenblott et al. 2018.
26. Horowitz 2020.
27. Wenger et al. 2020.

NOTES TO CHAPTER 3   203

28. Koskimaa & Raunio 2020.
29. Fuerth & Faber 2011.
30. Alongside these national efforts, foresight is also becoming the subject of international cooperation and collaboration. In 2020, the EU announced the establishment of an EU-wide foresight network. Each member state is responsible for appointing a ministerial-level representative, collectively termed ministers for the future. Beyond Europe, the OECD Structural Reform Support Division has taken up the objective of supporting governments to enhance their foresight capacity, and it is proposed the that United Nations take a more active role as well. Prítyi et al. 2021; United Nations 2022a.
31. Kolbert 2009.
32. Stockdale 2016.
33. Koskimaa & Raunio 2020.
34. Jay et al. 2007; Cashmore et al. 2004; McCullough 2017.
35. Boston 2016, p. 343.
36. McCullough 2017.
37. Boston 2016.
38. Stockdale 2016.
39. Nelson & Katzenstein 2020.
40. Sabel & Victor 2022.
41. Dewey 1927, p. 209.
42. Voß et al. 2006, p. 4.
43. Johnson 2018; Feindt & Weiland 2018.
44. Sabel & Zeitlin 2012; Sabel & Simon 2011.
45. Helimann 2018, p. 21.
46. Sabel & Victor 2022.
47. Geels et al. 2017; Victor et al. 2019.
48. Bartley 2010.
49. Levin et al. 2012, p. 123.
50. Farmer et al. 2019, p. 132.
51. Portions of the next several paragraphs are from Hale 2020.
52. Bernstein & Hoffmann 2018, 2019.
53. Geels et al. 2017; Zenghelis et al. 2018; Victor et al. 2019.
54. Hale & Urpelainen 2015.
55. Sawin 2001; Breetz et al. 2018.
56. Hoffmann 2011; Sabel & Victor 2022.
57. Biedenkopf et al. 2017.
58. Finnemore & Sikkink 1998, p. 895.
59. Bernstein & Hoffmann 2018.
60. Finnemore & Sikkink 1998.
61. Green 2018.
62. Urpelainen 2013; Meckling et al. 2015; Van der Ven et al. 2017; Breetz et al. 2018.
63. Colgan et al. 2020.
64. Downs et al. 1996.
65. Hale & Roger 2014; Betsill et al. 2015; Hale 2016, Slaughter 2017.

66. Bulkeley et al. 2014.
67. Hale 2016.
68. Sebenius 1991.
69. Bailey et al. 1997.
70. Bernstein & Hoffmann 2019.
71. Victor et al. 1998; Abbott 2017; Sabel & Victor 2017; Aldy 2018.
72. Roger et al. 2017.
73. Abbott 2017.
74. Urpelainen 2013; Young 2017.
75. Thomas 2001.
76. Kelley 2017.
77. Van Asselt 2016; Climate Action Tracker 2019.
78. Urpelainen 2009; Bromley-Trujillo et al. 2016; Cao & Ward 2017.
79. For assessments, see Keohane & Oppenheimer 2016, p. 10; Victor 2015; Bang et al. 2016; Barrett & Dannenberg 2016, 2017.

Chapter 4: The Long View

1. Karnein 2016; MacKenzie 2021a.
2. Keck & Sikkink 1998.
3. Boston 2016, p. 322.
4. Jasanoff 1990.
5. Rocco 2021.
6. Livingston & Rummukainen 2020.
7. Boykoff & Pearman 2019.
8. Kohl & Stenhouse 2021.
9. Hale et al. 2022.
10. Nash & Steurer 2019.
11. Averchenkova et al. 2021.
12. Averchenkova et al. 2021.
13. Averchenkova et al. 2021.
14. Krznaric 2020, p. 177.
15. For example, a left-leaning Australian government was able to create both a Climate Change Authority and a Climate Commission tasked with information provision, but strong political opposition limited their impact on policy outcomes. MacNeil 2021.
16. Shoham & Lamay 2006.
17. Boston 2016, p. 330.
18. Boston 2016, p. 331.
19. Beckman & Uggla 2016.
20. MacKenzie 2021a; Smith 2021.
21. Andersson & Keizer 2014.
22. Boston 2016, chap. 8.
23. Ilzetzki 2021.
24. Delaney & Dixon 2018.

NOTES TO CHAPTER 4 205

25. Pavone & Stiansen 2022.
26. Romelli 2022.
27. Woods 2006.
28. Romelli 2022.
29. Moravcsik 2000.
30. Meckling & Nahm 2018.
31. Carlson 2017.
32. Nichols 2010.
33. Vogel 2018.
34. Vogel 2018.
35. Meckling & Nahm 2018.
36. Setzer & Higham 2021.
37. Setzer & Benjamin 2020.
38. Quoted in Boston 2016, chap. 7.
39. Martinez & Winter 2021.
40. González-Ricoy 2016.
41. Araújo & Koessler 2021; Krznaric 2020.
42. Martinez & Winter 2021.
43. Grant & Keohane 2005.
44. Tremmel 2006.
45. Petryna 2022.
46. Orr 2016.
47. Young et al. 2017.
48. Sampson & Shi 2020.
49. Flammer & Bansal 2017.
50. Thomas 2019.
51. Hale et al. 2022.
52. Note, however, many environmental and social issues are not long problems. For example, it is important to remove forced labor from supply chains to help people in the present, but requiring companies to consider this will not alter their time horizons. So expanding what companies care about is orthogonal to extending the time period they care about.
53. Better Business Act Campaign 2022.
54. Costanza 2017.
55. Goldstein & Keohane 1993.
56. Krznaric 2020.
57. Keohane 2010.
58. Owens & Rayner 1999; Liftin 1994.
59. Boston 2016, p. 442.
60. Figueres & Rivett-Carnac 2020.
61. Boston, 2014, p. 27.
62. Asensio et al. 2022.
63. Asensio et al. 2022.
64. Asensio et al. 2022; Klinsky & Sagar 2022.
65. Hale 2020.

66. Busby 2022.
67. Anderson 2010.
68. Molthan-Hill et al. 2019; Monroe et al. 2019.
69. Rousell & Cutter-Mackenzie-Knowles 2020.
70. Darden & Mylonas 2015.
71. Devaney et al. 2020; Wells 2022.
72. Curato 2019; Westminster Foundation for Democracy 2021.
73. International Institute for Sustainable Development 2011, p. 29.
74. Dryzek 1996; Fishkin 2003, 2009; Bächtiger et al. 2018.
75. MacKenzie & Caluwaerts 2021.
76. Ghimire et al. 2021.
77. Willis et al. 2022; MacKenzie 2021a.
78. MacKenzie 2021a.
79. Niemeyer & Jennstål 2016.
80. Smith 2021. An interesting set of experiments in Japan assigned a subgroup of town hall meeting participants to explicitly imagine themselves at citizens in the future and found that such framing did not just affect the preferences of the subgroup but of the large deliberative body. Hara et al. 2019; Hiromitsu et al. 2021.
81. Kuntze & Fesenfeld 2021; MacKenzie & Caluwaerts 2021.
82. Wells et al. 2021.
83. Tarrow 1998.
84. Chenoweth 2021.
85. Culpepper 2010.
86. Hadden 2015.
87. Allan 2020.
88. De Moor et al. 2021; Fisher & Nasrin 2021.
89. Bell et al. 2021.
90. Kennedy & Johnson 2020.
91. Bell et al. 2021.
92. Bidadanure 2016.

Chapter 5: Endurance and Adaptability

1. Quoted in Tremmel 2017, p. 5.
2. Young 2017, p. 217.
3. Collins & Porras 1994, p. xiii. I thank Eric Beinhocker for the reference.
4. MacKenzie 2016.
5. Ginsburg et al. 2009.
6. Tsebelis 2002.
7. Young et al. 2017; Boston 2016, chap. 8.
8. Mazzucato 2020.
9. Young & Biermann 2017, pp. 122–123.
10. Heilmann 2011; Heilmann & Melton 2013.
11. Li 2019.

12. Kanie et al. 2019.
13. Kaplan 1983.
14. Christopher & Hood 2006; Verbeeten & Speklé 2015.
15. Biermann et al. 2017.
16. Young & Biermann 2017.
17. Kelley & Simmons 2014; Kelley 2017; Pintér et al. 2017.
18. Kelley & Simmons 2014; Morse 2022.
19. CDP 2023.
20. GFANZ 2022.
21. Hale 2017.
22. Andresen & Iguchi 2017.
23. McArthur & Rasmussen 2018.
24. Biermann et al. 2022.
25. United Nations 2022b.
26. Biermann et al. 2022, p. 796.
27. Christopher & Hood 2006; Pintér et al. 2017.
28. Kanie et al. 2019.
29. Zhao et al. 2020.
30. Kouroutakis 2017, p. 3.
31. Feindt & Weiland 2018.
32. Johnson 2018.
33. Voß et al. 2006, p. 19.
34. Kouroutakis 2017.
35. Tverberg 2017.
36. Ranchordás 2015.
37. Ranchordás 2015.
38. Bar-Siman-Tov 2018.
39. Cabinet Office 2015.
40. Kouroutakis & Ranchordás 2016.
41. Ansell & Adler 2019.
42. Manulak, 2022.
43. Manulak, 2022.
44. Zhao et al. 2020, p. 59.
45. Helimann 2018, p. 116.
46. Tremmel 2017.
47. Feindt & Weiland 2018, p. 666.
48. Colgan et al. 2020.
49. Ruggie 1993.
50. Boston 2016.
51. Weaver 1989.
52. Cashore & Howlett 2007.
53. NikkeiAsia 2014.
54. Auld et al. 2021, p. 716; emphases original.
55. Weaver 1989.

56. Botterill & Hayes 2012.
57. Kuijt & Finlayson 2009.
58. Wong et al. 1991.
59. Author's estimate based on Foreign Agricultural Service, 2023.
60. International Energy Agency 2022.
61. Aragon 2013.
62. Jacobs 2011.
63. OECD 2023.
64. Norwegian Ministry of Finance 2021.
65. Young 2020.
66. PIF 2023.
67. Weaver 1989.
68. Bernstein et al. 2013.
69. Philosophers have proposed versions of this idea, such as a Common Heritage Fund for Future Generations. Szabó 2016.

Chapter 6: Studying Long Problems

1. Bjornerud 2018, p. 16.
2. Wenger et al. 2020.
3. Quoted in Pierson 2004, p. 1.
4. Jacobs 2011.
5. Pierson 2004; Drezner 2021; Bernstein et al. 2000.
6. Fearon 1995.
7. Lars-Erik 2001.
8. Lipset & Rokkan 1967.
9. Hardin 1968; Hardin 1982; Barrett 2003; Sandler 2004.
10. Axelrod 1984.
11. Ostrom 2009; Jordan et al. 2018.
12. Aklin & Mildenberger 2020.
13. Hale & Urpelainen 2015.
14. Hale 2020.
15. Geels et al. 2017.
16. Busby & Urpelainen 2020.
17. Levin et al. 2012.
18. Bernstein & Hoffmann 2019.
19. Colgan et al. 2020.
20. Sabel & Victor 2022.
21. Gerlach & Rayner 1988.
22. Sabel & Victor, 2022.
23. See discussion in Ord 2020, p. 253; Beckerman & Hepburn 2007.
24. Lipscy, forthcoming; Jacobs 2011.
25. Simmons 2016.
26. Mani et al. 2013.

NOTES TO CHAPTER 6   209

27. Pierson 2000, 2004.
28. Darden & Mylonas 2015.
29. Jacobs & Weaver 2015.
30. Collier & Munck 2022.
31. Capoccia & Kelemen 2007, p. 348.
32. Ikenberry 2000.
33. Manulak, 2022.
34. Lipscy 2020.
35. Pierson 2004.
36. Finnemore & Sikkink 1998.
37. Stockdale 2016.
38. Jasanoff & Kim 2015.
39. Campbell 2004.
40. Bates et al. 1999; Greif 2006.
41. Simmons & Elkins 2004; Hale 2014.
42. Bates et al. 1999.
43. Axelrod 1984.
44. De Marchi & Page 2014.
45. For example, see Hovi et al. 2019.
46. Van Beek et al. 2020.
47. Although there is progress on this front. Peng et al. 2021; Moore et al. 2022.
48. Cash et al. 2003.
49. Amer et al. 2013.
50. Milkoreit 2017.
51. Finn & Wylie 2021.
52. Drezner 2021.
53. Liu 2014.
54. Nnadi & Carter 2021.
55. Farmer et al. 2019.
56. Bjornerud 2018.
57. Pelling & High 2005; Adger 2010; Aldrich et al. 2016.
58. Toffler 1970, p. 2.
59. In contrast, Greif models political institutions as a succession of interlinked games. Change emerges partially endogenously from the prior equilibrium, but no overarching ideas or drivers explain the general patterns of historical development. Greif 2006.
60. The scenarios here focus largely on material dynamics and drivers, but one could imagine many alternatives—for example, a pathway in which carbon emissions become widely seen as morally anathema. There are in fact current legal efforts to make "ecocide" a crime—shifting the politics of climate to resemble more those of, say, war crimes.
61. Adam 2010, p. 370.
62. Wells 1902, p. 392. Wells was not unaware of the difficulties of doing so but drew a helpful comparison to the past: "And the question arises how far this absolute ignorance of the future is a fixed and necessary condition of human life, and how far some application of intellectual methods may not attenuate even if does not absolutely set aside the veil between ourselves and

things to come. And I am venturing to suggest to you that along certain lines and with certain qualifications and limitations a working knowledge of things in the future is a possible and practicable thing. And in order to support this suggestion I would call your attention to certain facts about our knowledge of the past, and more particularly I would insist upon this, that about the past our range of absolute certainty is very limited indeed. About the past I would suggest we are inclined to overestimate our certainty, just as I think we incline to underestimate the certainties of the future" (p. 368).

63. Andersson 2018.
64. Tarsney 2022.
65. Tetlock 2005.
66. Tetlock 2005, p. xx.
67. Tetlock 2005, p. xx.
68. Adam 2010, p. 15.
69. Wenger et al. 2020, p. 229.
70. Adam 2010, p. 370.
71. Hellmann 2020.
72. Tetlock 2005, p. xxxiv.
73. Toffler 1970, p. 5.
74. Tarsney 2022.
75. Tetlock 2005.
76. Patomäki 2006, p. 3.
77. Bernstein et al. 2000, p. 43.
78. Sælen et al. 2020.
79. Castro et al. 2020.
80. Lamperti et al. 2018.
81. Barrett & Dannenberg 2012.
82. Bernstein et al. 2000; Levin et al. 2012.
83. Bernstein et al. 2000, p. 70.
84. Bernstein et al. 2000, p. 59.

Chapter 7: Governing Time

1. Quoted in MacKenzie 2021a, p. 200.
2. Colgan et al. 2020.
3. For example, the School of International Futures has developed a methodology for assessing intergenerational fairness in policy design. School of International Futures 2022. The United Kingdom's All Party Parliamentary Group on Future Generations has created a Future Check program to assess proposed legislation in the United Kingdom.
4. Dubash 2021.
5. Broz & Maliniak 2010.
6. As Toffler put it in his 1970 bestseller *Future Shock*, "To master change, we shall therefore need both a clarification of important long-range social goals *and* a democratization of the way in which we arrive at them. And this means nothing less than the next political revolution in the techno-societies—a breathtaking affirmation of popular democracy" (p. 477).

7. Paterson et al. 2022.
8. This entire section draws on Hale 2019.
9. Colgan et al. 2020.
10. Biermann 2014; Biermann et al. 2012; Young & Biermann 2017.
11. Caney 2022a.
12. Tremmel 2018; Boston 2016; Dirth 2019; Steven & Francruz 2021; Tonn 1996; Wenger et al. 2020; MacKenzie 2021a; Caney 2022a; Hale et al. 2023; Jasanoff 1990.
13. Beck 1992; Andersson 2018; Adam 2010.
14. Vickers 1970, p. 76.

Appendix 1: Why We Face More Long Problems Now

1. Global Footprint Network 2020.
2. Roser 2021.
3. Steffen et al. 2004; Steffen et al. 2015.
4. Steffen et al. 2004; Steffen et al. 2015.
5. Global Carbon Project 2020.
6. Butchart et al. 2010; WWF 2020; IPBES 2019.
7. Bjornerud 2018.
8. Rockström et al. 2009.
9. Frankopan 2023.
10. Green & Hale 2017.
11. Biermann 2014.
12. Delanty & Mota 2017; Bornemann & Strassheim 2019; Muiderman et al. 2020.
13. Nowotny 1994.
14. See also Porter & Stockdale 2016.
15. For a typology of timescales across human and nonhuman systems, see Fraser 1999.
16. Bjornerud 2018, p. 7.
17. Bjornerud 2018, p. 8.
18. Edwards 2010.
19. IPCC 2014.
20. Bolin 2007.
21. Edwards 2010, p. 434.
22. Andersson & Rindzevičiūtė 2015; Andersson 2018; Wenger et al. 2020.
23. Andersson 2018, p. 3.
24. Warde & Sörlin 2015.
25. Ogburn 1923; Toffler 1970; Beck 1992; Rosa 2013.
26. Nowotny 1994.
27. For more on such risks, see Rees 2003; Ord 2020.
28. Beck 1992.
29. Beck 1992, p. 48.
30. Beck 1992, p. 187.
31. Boston, 2016, p. 7.
32. United Nations Secretary-General 2013, p. 2.

33. Frederick et al. 2002.
34. Krznaric 2020; Ord 2020.
35. Held 1995, 2004; Caney 2005.
36. Broome 2012; Shue 2014.
37. Krznaric 2020.
38. Ord 2020, p. 45.
39. Araújo & Koessler 2021.
40. United Nations Secretary-General 2013.

Appendix 2: The Role of Acceleration

1. Rosa 2013.
2. One influential definition of globalization notes that "velocity" is a key dimension of increasing global interconnections. See Held et al. 1999.
3. Toffler 1970, p. 1.
4. Ogburn 1923.
5. Huntington 1968.
6. Marx & Engels [1848] 1967, p. 224. Quoted in Adam 2010, p. 367.
7. Fawcett 2018.
8. Schedler & Santiso 1998.
9. Vickers 1970, p. 93.
10. See Rosa 2013, p. 262, fig. 11.1 "Paradoxes of Political Time."
11. Goetz 2014.
12. Ogburn 1923, p. 199.
13. Rosa 2013, p. 259.

REFERENCES

Abbott, K. W. (2012). "The Transnational Regime Complex for Climate Change." *Environment and Planning C: Government and Policy* **30**(4): 571–590.

Abbott, K. W. (2017). "Orchestrating Experimentation in Non-state Environmental Commitments." *Environmental Politics* **26**(4): 738–763.

Abbott, K. W., J. Green, and R. O. Keohane (2016). "Organizational Ecology and Institutional Change in Global Governance." *International Organization* **70**(2): 247–277.

Adam, B. (2010). "History of the Future: Paradoxes and Challenges." *Rethinking History* **14**(3): 361–378.

Adger, W. N. (2010). "Social Capital, Collective Action, and Adaptation to Climate Change." In *Der Klimawandel: Sozialwissenschaftliche Perspektiven*, ed. M. Voss, 327–345. Wiesbaden: VS Verlag für Sozialwissenschaften.

Aklin, M., and M. Mildenberger. (2020). "Prisoners of the Wrong Dilemma: Why Distributive Conflict, Not Collective Action, Characterizes the Politics of Climate Change." *Global Environmental Politics* **20**(4): 4–27.

Aldrich, D. P., C. M. Page-Tan, and C. J. Paul. (2016). *Social Capital and Climate Change Adaptation*. Oxford: Oxford University Press.

Aldy, J. E. (2018). "Policy Surveillance: Its Role in Monitoring, Reporting, Evaluating and Learning." In *Governing Climate Change: Polycentricity in Action*, ed. A. Jordan, D. Huitema, H. van Asselt, and J. Forster, 210–227. Cambridge: Cambridge University Press.

Allan, J. I. (2020). *The New Climate Activism: NGO Authority and Participation in Climate Change Governance*. Toronto: University of Toronto Press.

Allison, G. T., and M. H. Halperin (1972). "Bureaucratic Politics: A Paradigm and Some Policy Implications." *World Politics* **24**: 40–79.

Amer, M., T. U. Daim, and A. Jetter (2013). "A Review of Scenario Planning." *Futures* **46**: 23–40.

Anderson, A. (2010). *Combating Climate Change through Quality Education*. Washington, DC: Brookings Institution.

Andersson, J. (2018). *The Future of the World: Futurology, Futurists, and the Struggle for the Post Cold War Imagination*. Oxford: Oxford University Press.

Andersson, J., and A.-G. Keizer (2014). "Governing the Future: Science, Policy and Public Participation in the Construction of the Long Term in the Netherlands and Sweden." *History and Technology* **30**(1–2): 104–122.

Andersson, J., and E. Rindzevičiūtė (2015). *The Struggle for the Long-Term in Transnational Science and Politics: Forging the Future*. London: Routledge.

Andresen, S., and M. Iguchi (2017). "Lessons from the Health-Related Millennium Development Goals." In *Governing through Goals: Sustainable Development Goals as Governance Innovation*, ed. N. Kanie and F. Biermann, 165–186. Cambridge, MA: MIT Press.

Ansell, B., and D. Adler (2019). "Brexit and the Politics of Housing in Britain." *Political Quarterly* **90**(S2): 105–116.

Aragon, C. T. (2013). "The United Nations Must Manage a Global Food Reserve." *UN Chronicle*. Retrieved June 29, 2023, from https://www.un.org/en/chronicle/article/united-nations-must-manage-global-food-reserve.

Araújo, R., and L. Koessler (2021). "The Rise of the Constitutional Protection of Future Generations." Legal Priorities Project Working Paper 7–2021.

Asensio, J. C., D. Blaquier, and J. Sedemund (2022). "Strengthening Capacity for Climate Action in Developing Countries." OECD Development Co-operation Working Paper No. 106.

Auld, G., S. Bernstein, B. Cashore, and K. Levin (2021). "Managing Pandemics as Super Wicked Problems: Lessons from, and for, COVID-19 and the Climate Crisis." *Policy Sciences* **54**(4): 707–728.

Averchenkova, A., S. Fankhauser, and J. J. Finnegan (2021). "The Influence of Climate Change Advisory Bodies on Political Debates: Evidence from the UK Committee on Climate Change." *Climate Policy* **21**(9): 1218–1233.

Axelrod, R. (1984). *The Evolution of Cooperation*. New York: Basic Books.

Bächtiger, A., J. S. Dryzek, J. Mansbridge, and M. E. Warren (2018). *The Oxford Handbook of Deliberative Democracy*. Oxford: Oxford University Press.

Bailey, M. A., J. L. Goldstein, and B. R. Weingast (1997). "The Institutional Roots of American Trade Policy: Politics, Coalitions, and International Trade." *World Politics* **49**(3): 309–338.

Bang, G., J. Hovi, and T. Skodvin (2016). "The Paris Agreement: Short-Term and Long-Term Effectiveness." *Politics and Governance* **4**(3): 209–218.

Bar-Siman-Tov, I. (2018). "Temporary Legislation, Better Regulation, and Experimentalist Governance: An Empirical Study." *Regulation and Governance* **12**(2): 192–219.

Barrett, S. (2003). *Environment and Statecraft*. Oxford: Oxford University Press.

Barrett, S., and A. Dannenberg (2012). "Climate Negotiations under Scientific Uncertainty." *Proceedings of the National Academy of Sciences* **109**(43): 17372–17376.

Barrett, S., and A. Dannenberg (2016). "An Experimental Investigation into 'Pledge and Review' in Climate Negotiations." *Climatic Change* **138**: 339–351.

Barrett, S., and A. Dannenberg (2017). "Tipping versus Cooperating to Supply a Public Good." *Journal of the European Economic Association* **15**(4): 910–941.

Bartley, T. (2010). "Transnational Private Regulation in Practice: The Limits of Forest and Labor Standards Certification in Indonesia." *Business and Politics* **12**(3): 1–34.

Bates, R. H., A. Greif, M. Levi, J.-L. Rosenthal, and B. R. Weingast (1999). *Analytic Narratives*. Princeton, NJ: Princeton University Press.

Beck, U. (1992). *Risk Society: Towards a New Modernity*. Thousand Oaks: Sage.

Beckerman, W., and C. Hepburn (2007). "Ethics and the Discount Rate in the Stern Review on the Economics of Climate Change." *World Economics* **8**(1): 187–210.

Beckman, L., and F. Uggla (2016). "An Ombudsman for Future Generations: Legitimate and Effective?" In *Institutions for Future Generations*, ed. I. González-Ricoy and A. Gosseries, 117–134. Oxford: Oxford University Press.

Bell, J., J. Poushter, M. Fagan, and C. Huang (2021). *In Response to Climate Change, Citizens in Advanced Economies Are Willing to Alter How They Live and Work*. Washington, DC: Pew Research Center.

Bernstein, S., and M. Hoffmann (2018). "The Politics of Decarbonization and the Catalytic Impact of Subnational Climate Experiments." *Policy Sciences* 51(2): 189–211.

Bernstein, S., and M. Hoffmann (2019). "Climate Politics, Metaphors and the Fractal Carbon Trap." *Nature Climate Change* 9(12): 919–925.

Bernstein, S., R. N. Lebow, J. G. Stein, and S. Weber (2000). "God Gave Physics the Easy Problems: Adapting Social Science to an Unpredictable World." *European Journal of International Relations* 6(1): 43–76.

Bernstein, S., J. Lerner, and A. Schoar (2013). "The Investment Strategies of Sovereign Wealth Funds." *Journal of Economic Perspectives* 27(2): 219–238.

Betsill, M., N. K. Dubash, M. Paterson, H. van Asselt, A. Vihma, and H. Winkler (2015). "Building Productive Links between the UNFCCC and the Broader Global Climate Governance Landscape." *Global Environmental Politics* 15(2): 1–10.

Better Business Act Campaign. (2022). *Is Legislation the Best Way to Achieve Stakeholder Capitalism?* London: Better Business Act Campaign.

Bidadanure, J. (2016). "Youth Quotas, Diversity, and Long-Termism: Can Young People Act as Proxies for Future Generations?" *Institutions for Future Generations*, ed. I. González-Ricoy and A. Gosseries, 266–281. Oxford: Oxford University Press.

Biedenkopf, K., P. Müller, P. Slominski, and J. Wettestad (2017). "A Global Turn to Greenhouse Gas Emissions Trading? Experiments, Actors, and Diffusion." *Global Environmental Politics* 17(3): 1–11.

Biermann, F. (2014). *Earth System Governance: World Politics in the Anthropocene*. Cambridge, MA: MIT Press.

Biermann, F., K. Abbott, S. Andresen, K. Bäckstrand, S. Bernstein, M. M. Betsill, H. Bulkeley, et al. (2012). "Navigating the Anthropocene: Improving Earth System Governance." *Science* 335(6074): 1306–1307.

Biermann, F., T. Hickmann, C.-A. Sénit, M. Beisheim, S. Bernstein, P. Chasek, L. Grob, et al. (2022). "Scientific Evidence on the Political Impact of the Sustainable Development Goals." *Nature Sustainability* 5: 795–800.

Biermann, F., N. Kanie, and R. E. Kim (2017). "Global Governance by Goal-Setting: The Novel Approach of the UN Sustainable Development Goals." *Current Opinion in Environmental Sustainability* 26–27: 26–31.

Bjornerud, M. (2018). *Timefulness: How Thinking like a Geologist Can Help Save the World*. Princeton, NJ: Princeton University Press.

Bolin, B. (2007). *A History of the Science and Politics of Climate Change: The Role of the Intergovernmental Panel on Climate Change*. Cambridge: Cambridge University Press.

Bornemann, B., and H. Strassheim (2019). "Governing Time for Sustainability: Analyzing the Temporal Implications of Sustainability Governance." *Sustainability Science* 14(4): 1001–1013.

Boston, J. (2014). "Governing for the Future: How to Bring the Long-Term into Short-Term Political Focus." Paper prepared for a seminar at the Centre for Environmental Policy, School of Public Affairs, American University, Washington, DC. Retrieved August 6, 2023, from https://www.american.edu/spa/cep/upload/jonathan-boston-lecture-american-university.pdf.

Boston, J. (2016). *Governing for the Future: Designing Democratic Institutions for a Better Tomorrow*. Bingley, UK: Emerald Group Publishing.

Botterill, L. C., and M. J. Hayes (2012). "Drought Triggers and Declarations: Science and Policy Considerations for Drought Risk Management." *Natural Hazards* **64**(1): 139–151.

Boykoff, M., and O. Pearman (2019). "Now or Never: How Media Coverage of the IPCC Special Report on 1.5°C Shaped Climate-Action Deadlines." *One Earth* **1**(3): 285–288.

Breetz, H., M. Mildenberger, and L. Stokes (2018). "The Political Logics of Clean Energy Transitions." *Business and Politics* **20**(4): 492–522.

Bromley-Trujillo, R., J. S. Butler, J. Poe, and W. Davis (2016). "The Spreading of Innovation: State Adoptions of Energy and Climate Policy." *Review of Policy Research* **33**(5): 544–565.

Broome, J. (2012). *Climate Matters: Ethics in a Warming World*. New York: Norton.

Broz, J. L., and D. Maliniak (2010). "Malapportionment, Gasoline Taxes, and Climate Change." Paper presented at the Annual Meeting of the American Political Science Association, Washington, DC, September 2–5.

Bulkeley, H., L. B. Andonova, M. Betsill, D. Compagnon, T. Hale, M. Hoffmann, P. Newell, M. Paterson, C. Roger, and S. D. VanDeveer (2014). *Transnational Climate Change Governance*. Cambridge: Cambridge University Press.

Busby, J. W. (2022). *States and Nature: The Effects of Climate Change on Security*. Cambridge: Cambridge University Press.

Busby, J. W., and J. Urpelainen (2020). "Following the Leaders? How to Restore Progress in Global Climate Governance." *Global Environmental Politics* **20**(4): 99–121.

Butchart, S. H. M., M. Walpole, B. Collen, A. van Strien, J. P. W. Scharlemann, R. E. A. Almond, J. E. M. Baillie, et al. (2010). "Global Biodiversity: Indicators of Recent Declines." *Science* **328**(5982): 1164–1168.

Cabinet Office (2015). *Better Regulation Framework Manual: Practical Guidance for UK Government Officials*. London: UK Cabinet Office.

Campbell, J. L. (2004). *Institutional Change and Globalization*. Princeton, NJ: Princeton University Press.

Caney, S. (2005). *Justice beyond Borders: A Global Political Theory*. Oxford: Oxford University Press.

Caney, S. (2022a). "Global Climate Governance, Short-Termism, and the Vulnerability of Future Generations." *Ethics and International Affairs* **36**(2): 137–155.

Caney, S. (2022b). "Global Climate Governance, Short-Termism, and the Vulnerability of Future Generations." Roundtable: Vulnerable Communities, Future Generations, and Political Representation in Climate Policy and Practice.

Cao, X., and H. Ward (2017). "Transnational Climate Governance Networks and Domestic Regulatory Action." *International Interactions* **43**(1): 76–102.

Capoccia, G., and R. D. Kelemen (2007). "The Study of Critical Junctures: Theory, Narrative, and Counterfactuals in Historical Institutionalism." *World Politics* **59**(3): 341–369.

Carlson, A. E. (2017). "Regulatory Capacity and State Environmental Leadership: California's Climate Policy." *Fordham Environmental Law Review* **24**(1): 63–86.

Cash, D. W., W. C. Clark, F. Alcock, N. M. Dickson, N. Eckley, D. H. Guston, J. Jäger, and R. B. Mitchell (2003). "Knowledge Systems for Sustainable Development." *Proceedings of the National Academy of Sciences* **100**(14): 8086.

Cashmore, M., R. Gwilliam, R. Morgan, D. Cobb, and A. Bond (2004). "The Interminable Issue of Effectiveness: Substantive Purposes, Outcomes and Research Challenges in the Advancement of Environmental Impact Assessment Theory." *Impact Assessment and Project Appraisal* **22**(4): 295–310.

Cashore, B., and M. Howlett (2007). "Punctuating Which Equilibrium? Understanding Thermostatic Policy Dynamics in Pacific Northwest Forestry." *American Journal of Political Science* **51**(3): 532–551.

Castro, J., S. Drews, F. Exadaktylos, J. Foramitti, F. Klein, T. Konc, I. Savin, and J. van den Bergh (2020). "A Review of Agent-Based Modeling of Climate-Energy Policy." *WIREs Climate Change* **11**(4): e647.

CDP (2023). "The CDP-WWF Temperature Rating Methodology." Retrieved April 30, 2023, from https://www.cdp.net/en/investor/temperature-ratings/cdp-wwf-temperature-ratings-methodology.

Chenoweth, E. (2021). *Civil Resistance: What Everyone Needs to Know*. Oxford: Oxford University Press.

Chermack, T. J., S. A. Lynham, and W. E. A. Ruona (2001). "A Review of Scenario Planning Literature." *Futures Research Quarterly* 17: 7–31.

Christopher, H., and C. Hood (2006). "Gaming in Targetworld: The Targets Approach to Managing British Public Services." *Public Administration Review* **66**(4): 515–521.

Climate Action Tracker (2019). *Warming Projections Global Update*. Cologne: NewClimate Institute.

Coase, R. H. (1960). "The Problem of Social Cost." *Journal of Law and Economics* 3: 1–44.

Colgan, J. D., J. F. Green, and T. N. Hale (2020). "Asset Revaluation and the Existential Politics of Climate Change." *International Organization* 75(2): 586–610.

Collier, D., and G. L. Munck (2022). *Critical Junctures and Historical Legacies: Insights and Methods for Comparative Social Science*. Lanham, MD: Rowman & Littlefield.

Collingridge, D. (1980). *The Social Control of Technology*. New York: St. Martin's.

Collins, J., and J. I. Porras (1994). *Built to Last: Successful Habits of Visionary Companies*. New York: Penguin Random House.

Costanza, R. (2017). "Toward a Sustainable Wellbeing Economy." *Solutions Journal* Special Issue.

Culpepper, P. D. (2010). *Quiet Politics and Business Power: Corporate Control in Europe and Japan*. Cambridge: Cambridge University Press.

Curato, N. (2019). *Democracy in a Time of Misery: From Spectacular Tragedies to Deliberative Action*. Oxford: Oxford University Press.

Darden, K., and H. Mylonas (2015). "Threats to Territorial Integrity, National Mass Schooling, and Linguistic Commonality." *Comparative Political Studies* **49**(11): 1446–1479.

de Marchi, S., and S. E. Page (2014). "Agent-Based Models." *Annual Review of Political Science* 17: 1–20.

de Moor, J., M. De Vydt, K. Uba, and M. Wahlström (2021). "New Kids on the Block: Taking Stock of the Recent Cycle of Climate Activism." *Social Movement Studies* **20**(5): 619–625.

Delaney, E. F., and R. Dixon (2018). *Comparative Judicial Review*. Cheltenham, UK: Edward Elgar.

Delanty, G., and A. Mota (2017). "Governing the Anthropocene: Agency, Governance, Knowledge." *European Journal of Social Theory* **20**(1): 9–38.

Devaney, L., D. Torney, P. Brereton, and M. Coleman (2020). "Ireland's Citizens' Assembly on Climate Change: Lessons for Deliberative Public Engagement and Communication." *Environmental Communication* **14**(2): 141–146.

Dewey, J. (1927). *The Public and Its Problems*. New York: Alan Swallow.

Dirth, E. (2019). *The Futuring Tool: A Toolkit for Responding to the Demands of the Fridays for Future Movement*. Potsdam: Institute for Advanced Sustainability Studies.

Downs, G., D. M. Rocke, and P. N. Barsoom (1996). "Is the Good News about Compliance Good News about Cooperation?" *International Organization* **50**(2): 379–406.

Drescher, S., and P. C. Emmer (2010). *Who Abolished Slavery? Slave Revolts and Abolitionism: A Debate with João Pedro Marques*. New York: Berghahn.

Dreyer, I., and G. Stang (2013). "Foresight in Governments: Practices and Trends around the World." *Yearbook of European Security*. Paris: EUISS.

Drezner, D. (2021). "Power and International Relations: A Temporal View." *European Journal of International Relations* **27**(1): 29–52.

Dryzek, J. S. (1996). *Deliberative Global Politics: Discourse and Democracy in a Divided World*. Cambridge: Polity.

Dubash, N. (2021). "Varieties of Climate Governance: The Emergence and Functioning of Climate Institutions." *Environmental Politics* **30**(Sup1): 1–25.

Edwards, P. N. (2010). *A Vast Machine: Computer Models, Climate Data, and the Politics of Global Warming*. Cambridge, MA: MIT Press.

Farmer, J. D., C. Hepburn, M. C. Ives, T. Hale, T. Wetzer, P. Mealy, R. Rafaty, S. Srivastav, and R. Way (2019). "Sensitive Intervention Points in the Post-carbon Transition." *Science* **364**(6436): 132–134.

Fawcett, P. (2018). "Doing Democracy and Governance in the Fast Lane? Towards a 'Politics of Time' in an Accelerated Polity." *Australian Journal of Political Science* **53**(4): 548–564.

Fearon, J. D. (1995). "Rationalist Explanations for War." *International Organization* **49**(3): 379–414.

Feindt, P. H., and S. Weiland (2018). "Reflexive Governance: Exploring the Concept and Assessing Its Critical Potential for Sustainable Development. Introduction to the Special Issue." *Journal of Environmental Policy & Planning* **20**(6): 661–674.

Figueres, C., and T. Rivett-Carnac (2020). *The Future We Choose*. New York: Manila Press.

Finn, E., and R. Wylie (2021). "Collaborative Imagination: A Methodological Approach." *Futures* **132**: 102788.

Finnegan, J. J. (2022). "Institutions, Climate Change, and the Foundations of Long-Term Policymaking." *Comparative Political Studies* **55**(7): 1198–1235.

Finnemore, M., and K. Sikkink (1998). "International Norm Dynamics and Political Change." *International Organization* **52**(4): 887–917.

Fisher, D. R., and S. Nasrin (2021). "Climate Activism and Its Effects." *WIREs Climate Change* **12**(1): e683.

Fishkin, J. S. (2003). "Consulting the Public through Deliberative Polling." *Journal of Policy Analysis and Management* **22**(1): 128–133.

Fishkin, J. S. (2009). *When the People Speak: Deliberative Democracy and Public Consultation.* Oxford: Oxford University Press.

Flammer, C., and P. Bansal (2017). "Does a Long-Term Orientation Create Value? Evidence from a Regression Discontinuity." *Strategic Management Journal* **38**(9): 1827–1847.

Flood, E. (2013). "Solyndra: Rhetoric and Reality in a Partisan Age Feature." *DttP: Documents to the People* **41**(4): 41–46.

Foreign Agricultural Service (2023). *Production, Supply, and Distribution.* Washington, DC: US Department of Agriculture.

Frankopan, P. (2023). *The Earth Transformed.* London: Bloomsbury.

Fraser, J. T. (1999). *Time, Conflict, and Human Values.* Urbana: University of Illinois Press.

Frederick, S., G. Loewenstein, and T. O'Donoghue (2002). "Time Discounting and Time Preference: A Critical Review." *Journal of Economic Literature* **40**(2): 351–401.

Fuerth, L., and E. M. H. Faber (2011). *Anticipatory Governance Practical Upgrades: Equipping the Executive Branch to Cope with Increasing Speed and Complexity of Major Challenges.* N.p.: Forward Engagement Project.

Galik, C. S. (2020). "A Continuing Need to Revisit BECCS and Its Potential." *Nature Climate Change* **10**(1): 2–3.

Gates, B. (2021). *How to Avoid a Climate Disaster: The Solutions We Have and the Breakthroughs We Need.* New York: Allen Lane.

Gazmararian, A., and H. Milner (2022). "Preference Updating under Uncertainty: Evidence from Responses to Global Warming." Unpublished manuscript.

Geels, F. W., B. K. Sovacool, T. Schwanen, and S. Sorrell (2017). "Sociotechnical Transitions for Deep Decarbonization." *Science* **357**(6357): 1242.

Gerlach, L., and S. Rayner (1988). *Managing Global Climate Change: A View from the Social and Decision Sciences.* Oak Ridge, TN: Oak Ridge National Laboratory.

Ghimire, R., N. Anbar, and N. B. Chhetri (2021). "The Impact of Public Deliberation on Climate Change Opinions among U.S. Citizens." *Frontiers in Political Science* **3**. Retrieved June 26, 2023, from https://www.frontiersin.org/articles/10.3389/fpos.2021.606829/full.

Ginsburg, T., Z. Elkins, and J. Melton (2009). *The Lifespan of Written Constitutions.* Chicago: University of Chicago Press.

Glasgow Financial Alliance for Net Zero (GFANZ) (2022). *The Glasgow Financial Alliance for Net Zero 2022 Progress Report.* London: GFANZ.

Global Carbon Project (2020). *Supplemental Data of Global Carbon Budget 2020 (Version 1.0)* [Data set]. N.p.: Global Carbon Project.

Global Footprint Network (2020). "Past Earth Overshoot Days." Retrieved April 2, 2021, from https://www.overshootday.org/newsroom/past-earth-overshoot-days/.

Goetz, K. H. (2014). "A Question of Time: Responsive and Responsible Democratic Politics." *West European Politics* **37**(2): 379–399.

Goldstein, J., and R. O. Keohane (1993). *Ideas and Foreign Policy Beliefs, Institutions, and Political Change.* Ithaca: Cornell University Press.

Goldstein, J. L., and R. Gulotty (2021). "America and the Trade Regime: What Went Wrong?" *International Organization* **75**(2): 524–557.

González-Ricoy, I. (2016). "Constitutionalizing Intergenerational Provisions." In *Institutions for Future Generations,* ed. I. González-Ricoy and A. Gosseries, 170–183. Oxford: Oxford University Press.

González-Ricoy, I., and A. Gosseries, eds. (2016). *Institutions for Future Generations*. Oxford: Oxford University Press.

Grant, R. W., and R. O. Keohane (2005). "Accountability and Abuses of Power in World Politics." *American Political Science Review* **99**(1): 29–43.

Grantham Research Institute on Climate Change and the Environment (2021). Climate Change Laws of the World Database.

Green, F. (2018). "Anti-Fossil Fuel Norms." *Climate Change* **150**: 103–116.

Green, J. F., J. Hadden, T. Hale, and P. Mahdavi (2022). "Transition, Hedge, or Resist? Understanding Political and Economic Behavior toward Decarbonization in the Oil and Gas Industry." *Review of International Political Economy* **29**(6): 2036–2206.

Green, J. F., and T. N. Hale (2017). "Reversing the Marginalization of Global Environmental Politics in International Relations: An Opportunity for the Discipline." *PS: Political Science and Politics* **50**(2): 473–479.

Greenblott, J. M., T. O'Farrell, R. Olson, and B. Burchard (2018). "Strategic Foresight in the Federal Government: A Survey of Methods, Resources, and Institutional Arrangements." *World Futures Review* **11**(3): 245–266.

Greif, A. (2006). *Institutions and the Path to the Modern Economy: Lessons from Medieval Trade*. Cambridge: Cambridge University Press.

Guston, D. H. (2014). "Understanding 'Anticipatory Governance.'" *Social Studies of Science* **44**(2): 218–242.

Hadden, J. (2015). *Networks in Contestation: The Divisive Politics of Climate Change*. Cambridge: Cambridge University Press.

Hale, T. (2014). "The Rule of Law in the Global Economy: Explaining Intergovernmental Backing for Private Commercial Tribunals." *European Journal of International Relations* **21**(3): 483–512.

Hale, T. (2016). "'All Hands on Deck': The Paris Agreement and Nonstate Climate Action." *Global Environmental Politics* **16**(3): 12–22.

Hale, T. (2017). "Climate Change: From Gridlock to Catalyst." In *Beyond Gridlock*, ed. T. Hale and D. Held, 184–204. Cambridge: Polity.

Hale, T. (2020). "Catalytic Cooperation." *Global Environmental Politics* **20**(4): 73–98.

Hale, T., A. J. Hale, B. Kira, A. Petherick, T. Phillips, D. Sridhar, R. N. Thompson, S. Webster, and N. Angrist (2020). "Global Assessment of the Relationship between Government Response Measures and COVID-19 Deaths." *medRxiv*: 2020.2007.2004.20145334.

Hale, T., and D. Held, eds. (2011). *Handbook of Transnational Governance: Institutions and Innovations*. Cambridge: Polity.

Hale, T., D. Held, and K. Young (2013). *Gridlock: Why Multilateralism Is Failing When We Need It Most*. Cambridge: Polity.

Hale, T., and C. Roger (2014). "Orchestration and Transnational Climate Governance." *Review of International Organizations* **9**(1): 59–82.

Hale, T., S. M. Smith, R. Black, K. Cullen, B. Fay, J. Lang, and S. Mahmood (2022). "Assessing the Rapidly-Emerging Landscape of Net Zero Targets." *Climate Policy* **22**(1): 18–29.

Hale, T., and J. Urpelainen (2015). "When and How Can Unilateral Policies Promote the International Diffusion of Environmental Policies and Clean Technology?" *Journal of Theoretical Politics* **27**(2): 177–205.

Hale, T. N. (2019). "Eleven Years to Save the Planet? The Closing Window for Climate Mitigation." *Oxford Government Review* (4): 32–33.

Hale, T. N., F. Moorhouse, T. Ord, and A.-M. Slaughter (2023). *Toward a Generation on Future Generations.* Oxford: Blavatnik School of Government.

Hara, K., R. Yoshioka, M. Kuroda, S. Kurimoto, and T. Saijo (2019). "Reconciling Intergenerational Conflicts with Imaginary Future Generations: Evidence from a Participatory Deliberation Practice in a Municipality in Japan." *Sustainability Science* 14(6): 1605–1619.

Hardin, G. (1968). "The Tragedy of the Commons." *Science* 162(3859): 1243–1248.

Hardin, R. (1982). *Collective Action.* Baltimore: Johns Hopkins University Press.

Heilmann, S. (2011). "Making Plans for Markets: Policy for the Long-Term in China." *Harvard Asia Quarterly* 13(2): 33–40.

Heilmann, S., and O. Melton (2013). "The Reinvention of Development Planning in China, 1993–2012." *Modern China* 39(6): 580–628.

Held, D. (1995). *Democracy and the Global Order.* Stanford: Stanford University Press.

Held, D. (2004). *Global Covenant: The Social Democratic Alternative to the Washington Consensus.* Cambridge: Polity.

Held, D., A. McGrew, D. Goldblatt, and J. Perraton (1999). *Global Transformations: Politics, Economics, and Culture.* Palo Alto: Stanford University Press.

Helg, A. (2019). *Slave No More: Self-Liberation before Abolitionism in the Americas.* Chapel Hill: University of North Carolina Press.

Helimann, S. (2018). *Red Swan: How Unorthodox Policy-Making Facilitated China's Rise.* New York: Columbia University Press.

Helleiner, E., S. Pagliari, and H. Zimmermann, eds. (2010). *Global Finance in Crisis: The Politics of International Regulatory Change.* London: Routledge.

Hellmann, G. (2020). "How to Know the Future—and the Past (and How to Not)." In *The Politics and Science of Prevision: Governing and Probing the Future,* ed. A. Wenger, U. Jasper, and M. Dunn Cavelty, 45–62. London: Routledge.

Hill, A. C., and L. Martinez-Diaz (2019). *Building a Resilient Tomorrow.* Oxford: Oxford University Press.

Hiromitsu, T., Y. Kitakaji, K. Hara, and T. Saijo (2021). "What Do People Say When They Become 'Future People'? Positioning Imaginary Future Generations (IFGs) in General Rules for Good Decision-Making." *Sustainability* 13(2): 6631.

Hoffmann, M. (2011). *Climate Governance at the Crossroads: Experimenting with a Global Response after Kyoto.* Oxford: Oxford University Press.

Hom, A. R. (2018). "Silent Order: The Temporal Turn in Critical International Relations." *Millennium* 46(3): 303–330.

Horowitz, M. C. (2020). "Future Thinking and Cognitive Distortions: Key Questions That Guide Forecasting Processes." In *The Politics and Science of Prevision: Governing and Probing the Future,* ed. A. Wenger, U. Jasper, and M. Dunn Cavelty, 63–72. London: Routledge.

Hovi, J., D. F. Sprinz, H. Sælen, and A. Underdal (2019). "The Club Approach: A Gateway to Effective Climate Co-operation?" *British Journal of Political Science* 49(3): 1071–1096.

Howe, P. D., J. R. Marlon, M. Mildenberger, and B. S. Shield (2019). "How Will Climate Change Shape Climate Opinion?" *Environmental Research Letters* 14(11): 113001.

Huntington, S. P. (1968). *Political Order in Changing Societies.* New Haven: Yale University Press.

Ikenberry, G. J. (2000). *After Victory: Institutions, Strategic Restraint, and the Rebuilding of Order after Major Wars.* Princeton, NJ: Princeton University Press.

Ilzetzki, E. (2021). Fiscal Rules in the European Monetary Union. *VoxEu,* June 10. Retrieved June 26, 2023, from https://cepr.org/voxeu/columns/fiscal-rules-european-monetary-union.

Intergovernmental Panel on Climate Change (IPCC) (2014). *Climate Change 2014: Mitigation of Climate Change. Contribution of Working Group III to the Fifth Assessment Report of the Intergovernmental Panel on Climate Change.* Cambridge: IPCC.

Intergovernmental Science-Policy Platform on Biodiversity and Ecosystem Services (IPBES) (2019). *Summary for Policymakers of the Global Assessment Report on Biodiversity and Ecosystem Services of the Intergovernmental Science-Policy Platform on Biodiversity and Ecosystem Services.* Bonn: IPBES Secretariat.

International Energy Agency (IEA) (2022). *IEA Member Countries to Make 60 Million Barrels of Oil Available Following Russia's Invasion of Ukraine.* Paris: IEA.

International Institute for Sustainable Development (2011). "Summary of the Durban Climate Change Conference: 28 November–11 December 2011." *Earth Negotiations Bulletin* **12**(534): 1–33.

Jacob, K., S. Weiland, J. Ferretti, D. Wascher, and D. Chodorowska (2011). *Integrating the Environment in Regulatory Impact Assessments.* Paris: OECD.

Jacobs, A. M. (2011). *Governing for the Long Term: Democracy and the Politics of Investment.* Cambridge: Cambridge University Press.

Jacobs, A. M. (2016). "Policy Making for the Long Term in Advanced Democracies." *Annual Review of Political Science* **19**: 433–454.

Jacobs, A. M., and J. S. Matthews (2012). "Why Do Citizens Discount the Future? Public Opinion and the Timing of Policy Consequences." *British Journal of Political Science* **42**(4): 903–935.

Jacobs, A. M., and R. K. Weaver (2015). "When Policies Undo Themselves: Self-Undermining Feedback as a Source of Policy Change." *Governance* **28**(4): 441–457.

Jasanoff, S. (1990). *The Fifth Branch: Science Advisers as Policymakers.* Cambridge, MA: Harvard University Press.

Jasanoff, S. (2020). "Imagined Worlds." In *The Politics and Science of Prevision: Governing and Probing the Future,* ed. A. Wenger, U. Jasper, and M. Dunn Cavelty, 27–44. London: Routledge.

Jasanoff, S., and S.-H. Kim (2015). *Dreamscapes of Modernity: Sociotechnical Imaginaries and the Fabrication of Power.* Chicago: University of Chicago Press.

Jay, S., C. Jones, P. Slinn, and C. Wood (2007). "Environmental Impact Assessment: Retrospect and Prospect." *Environmental Impact Assessment Review* **27**(4): 287–300.

Johnson, S. (2018). *Farsighted: How We Make the Decisions That Matter Most.* New York: Riverhead Books.

Jones, B. (1994). *Reconceiving Decision-Making in Democratic Politics: Attention, Choice, and Public Policy.* Chicago: University of Chicago Press.

Jordan, A., D. Huitema, H. van Asselt, and J. Forster (2018). *Governing Climate Change: Polycentricity in Action.* Cambridge: Cambridge University Press.

Jupille, J., W. Mattli, and D. Snidal (2013). *Institutional Choice and Global Commerce.* Cambridge: Cambridge University Press.

Kanie, N., and F. Biermann, eds. (2017). *Governing through Goals: Sustainable Development Goals as Governance Innovation.* Cambridge, MA: MIT Press.

Kanie, N., D. Griggs, O. Young, S. Waddell, P. Shrivastava, P. M. Haas, W. Broadgate, O. Gaffney, and C. Kőrösi (2019). "Rules to Goals: Emergence of New Governance Strategies for Sustainable Development." *Sustainability Science* 14(6): 1745–1749.

Kaplan, R. S. (1983). "Measuring Manufacturing Performance: A New Challenge for Managerial Accounting Research." *Accounting Review* 58(4): 686–705.

Karnein, A. (2016). "Can We Represent Future Generations?" In *Institutions for Future Generations,* ed. I. González-Ricoy and A. Gosseries, 83–97. Oxford: Oxford University Press.

Keck, M. E., and K. Sikkink (1998). *Activists beyond Borders.* Ithaca: Cornell University Press.

Kelley, C. P., S. Mohtadi, M. A. Cane, R. Seager, and Y. Kushnir (2015). "Climate Change in the Fertile Crescent and Implications of the Recent Syrian Drought." *Proceedings of the National Academy of Sciences* 112(11): 3241–3246.

Kelley, J. G. (2017). *Scorecard Diplomacy: Grading States to Influence Their Reputation and Behavior.* Cambridge: Cambridge University Press.

Kelley, J. G., and B. A. Simmons (2014). "Politics by Number: Indicators as Social Pressure in International Relations." *American Political Science Review* 59(1): 55–70.

Kendi, I. X. (2017). *Stamped from the Beginning: The Definitive History of Racist Ideas in America.* New York: Penguin.

Kennedy, B., and C. Johnson (2020). *More Americans See Climate Change as a Priority, but Democrats Are Much More Concerned than Republicans.* Washington, DC: Pew Research Center.

Keohane, N. O. (2010). *Thinking about Leadership.* Princeton, NJ: Princeton University Press.

Keohane, R. O., and M. Oppenheimer (2016). "Paris: Beyond the Climate Dead End through Pledge and Review?" *Politics and Governance* 4(3): 142–151.

Keohane, R. O., and D. G. Victor (2011). "The Regime Complex for Climate Change." *Perspectives on Politics* 9(1): 7–23.

Kim, T. A.-C. A., and R. Harrison (1999). *Self and Future Generations: An Intercultural Conversation.* Winwick, UK: White Horse Press.

Klinsky, S., and A. D. Sagar (2022). "The Why, What and How of Capacity Building: Some Explorations." *Climate Policy* 22(5): 549–556.

Kohl, P. A., and N. Stenhouse (2021). "12 Years Left: How a Climate Change Action Deadline Influences Perceptions and Engagement." *Environmental Communication* 15(7): 986–1000.

Kolbert, E. (2009). "Hosed: Is There a Quick Fix for the Climate?" *New Yorker,* November 16.

Koremenos, B. (2016). *The Continent of International Law: Explaining Agreement Design.* Cambridge: Cambridge University Press.

Koremenos, B., C. Lipson, and D. Snidal (2001). "The Rational Design of International Institutions." *International Organization* 55(4): 761–799.

Koskimaa, V., and T. Raunio (2020). "Encouraging a Longer Time Horizon: The Committee for the Future in the Finnish Eduskunta." *Journal of Legislative Studies* 26(2): 159–179.

Kouroutakis, A. (2017). *The Constitutional Value of Sunset Clauses: A Historical and Normative Analysis*. London: Routledge.

Kouroutakis, A., and S. Ranchordás (2016). "Snoozing Democracy: Sunset Clauses, Dejuridification, and Emergencies." *Minnesota Journal of International Law* 25(1): 29–78.

Krznaric, R. (2020). *The Good Ancestor: How to Think Long Term in a Short-Term World*. London: Penguin.

Kuijt, I., and B. Finlayson (2009). "Evidence for Food Storage and Predomestication Granaries 11,000 Years Ago in the Jordan Valley." *Proceedings of the National Academy of Sciences of the United States of America* 106(27): 10966–10970.

Kuntze, L., and L. P. Fesenfeld (2021). "Citizen Assemblies Can Enhance Political Feasibility of Ambitious Climate Policies." Retrieved June 26, 2023, from https://papers.ssrn.com/sol3/papers.cfm?abstract_id=3918532.

Lamperti, F., G. Dosi, M. Napoletano, A. Roventini, and A. Sapio (2018). "Faraway, So Close: Coupled Climate and Economic Dynamics in an Agent-Based Integrated Assessment Model." *Ecological Economics* 150: 315–339.

Lars-Erik, C. (2001). "Back to Kant: Reinterpreting the Democratic Peace as a Macrohistorical Learning Process." *American Political Science Review* 95(1): 15–31.

Lavery, S., and B. Donovan (2005). "Flood Risk Management in the Thames Estuary Looking Ahead 100 Years." *Philosophical Transactions of the Royal Society A: Mathematical, Physical and Engineering Sciences* 363(1831): 1455–1474.

Levin, K., B. Cashore, S. Bernstein, and G. Auld (2012). "Overcoming the Tragedy of Super Wicked Problems: Constraining Our Future Selves to Ameliorate Global Climate Change." *Policy Sciences* 45: 123–152.

Li, Y. (2019). "Bureaucracies Count: Environmental Governance through Goal-Setting and Mandate-Making in Contemporary China." *Environmental Sociology* 5(1): 12–22.

Libman, A. (2020). "Temporality of Authoritarian Regimes." In *The Oxford Handbook of Time and Politics*, ed. K. H. Goetz. Oxford: Oxford University Press.

Liftin, K. T. (1994). *Ozone Discourses*. New York: Columbia University Press.

Lindblom, C. E. (1959). "The Science of 'Muddling Through.'" *Public Administration Review* 19(2): 79–88.

Lipscy, P. Y. (2020). "COVID-19 and the Politics of Crisis." *International Organization* 74(S1): E98–E127.

Lipscy, P. Y. (Forthcoming). *The Institutional Politics of Energy and Climate Change*.

Lipset, S. M., and S. Rokkan (1967). "Cleavage Structures, Party Systems, and Voter Alignments: An Introduction." In *Party Systems and Voter Alignments: Cross-National Perspectives*, ed. S. M. Lipset and R. Rokkan, 3–64. New York: Free Press.

Liu, C. (2014). *The Three Body Problem*. New York: Tor Books.

Livingston, J. E., and M. Rummukainen (2020). "Taking Science by Surprise: The Knowledge Politics of the IPCC Special Report on 1.5 Degrees." *Environmental Science & Policy* 112: 10–16.

Lowi, T. J. (1964). "American Business, Public Policy, Case-Studies, and Political Theory." *World Politics* 16(4): 677–715.

Macaskill, W. (2022). *What We Owe the Future*. New York: Basic Books.

MacKenzie, M. K. (2016). "Institutional Design and Sources of Short-Termism." In *Institutions for Future Generations*, ed. I. González-Ricoy and A. Gosseries, 24–46. Oxford: Oxford University Press.

MacKenzie, M. K. (2021a). *Future Publics: Democracy, Deliberation, and Future-Regarding Collective Action*. Oxford: Oxford University Press.

MacKenzie, M. K. (2021b). "There Is No Such Thing as a Short-Term Issue." *Futures* **125**: 102652.

MacKenzie, M. K., and D. Caluwaerts (2021). "Paying for the Future: Deliberation and Support for Climate Action Policies." *Journal of Environmental Policy & Planning* **23**(3): 317–331.

MacNeil, R. (2021). "Swimming against the Current: Australian Climate Institutions and the Politics of Polarisation." *Environmental Politics* **30**(Sup1): 162–183.

Mani, A., S. Mullainathan, E. Shafir, and J. Zhao (2013). "Poverty Impedes Cognitive Function." *Science* **341**(6149): 976.

Manulak, M. W. (2022). *Change in Global Environmental Politics: Temporal Focal Points and the Reform of International Institutions*. Oxford: Oxford University Press.

Martinez, E., and C. Winter (2021). "Protecting Future Generations: A Global Survey of Legal Academics." Legal Priorities Project Working Paper 1–2021.

Martínez, E., and C. Winter (2023). "The Intuitive Appeal of Legal Protection for Future Generations." In *Essays on Longtermism*, ed. J. Barrett, D. Thorstad, and H. Greaves. Oxford: Oxford University Press. https://papers.ssrn.com/sol3/papers.cfm?abstract_id=4349899#:~:text=Here%20we%20present%20work%20from,laypeople%2C%20independent%20of%20demographic%20factors.

Mazzucato, M. (2020). *Mission Economy*. London: Allen Lane.

Mbaye, S., M. Moreno-Badia, and K. Chae (2021). "Global Debt Database: Methodology and Sources." IMF Working Paper.

McArthur, J. W., and K. Rasmussen (2018). "Change of Pace: Accelerations and Advances during the Millennium Development Goal Era." *World Development* **105**: 132–143.

McCullough, A. (2017). "Environmental Impact Assessments in Developing Countries: We Need to Talk about Politics." *Extractive Industries and Society* **4**(3): 448–452.

Meckling, J., N. Kelsey, E. Biber, and J. Zysman (2015). "Winning Coalitions for Climate Policy." *Science* **349**(6253): 1170.

Meckling, J., and J. Nahm (2018). "The Power of Process: State Capacity and Climate Policy." *Governance* **31**(4): 741–757.

Milkoreit, M. (2017). "Imaginary Politics: Climate Change and Making the Future." *Elementa: Science of the Anthropocene* **5**: 62.

Mitchell, R. B. (2006). "Problem Structure, Institutional Design, and the Relative Effectiveness of International Environmental Agreements." *Global Environmental Politics* **6**(3): 72–89.

Molthan-Hill, P., N. Worsfold, G. J. Nagy, W. Leal Filho, and M. Mifsud (2019). "Climate Change Education for Universities: A Conceptual Framework from an International Study." *Journal of Cleaner Production* **226**: 1092–1101.

Monroe, M. C., R. R. Plate, A. Oxarart, A. Bowers, and W. A. Chaves (2019). "Identifying Effective Climate Change Education Strategies: A Systematic Review of the Research." *Environmental Education Research* **25**(6): 791–812.

Moore, F. C., K. Lacasse, K. J. Mach, Y. A. Shin, L. J. Gross, and B. Beckage (2022). "Determinants of Emissions Pathways in the Coupled Climate–Social System." *Nature* **603**(7899): 103–111.

Moravcsik, A. (2000). "The Origins of Human Rights Regimes: Democratic Delegation in Postwar Europe." *International Organization* **54**(2): 217–252.

Morse, J. C. (2022). *The Bankers' Blacklist: Unofficial Market Enforcement and the Global Fight against Illicit Financing*. Ithaca: Cornell University Press.

Muiderman, K., A. Gupta, J. Vervoort, and F. Biermann (2020). "Four Approaches to Anticipatory Climate Governance: Different Conceptions of the Future and Implications for the Present." *WIREs Climate Change* **11**(6): e673.

Nash, S. L., and R. Steurer (2019). "Taking Stock of Climate Change Acts in Europe: Living Policy Processes or Symbolic Gestures?" *Climate Policy* **19**(8): 1052–1065.

Nelson, S. C., and P. J. Katzenstein (2020). "Crisis, What Crisis? Uncertainty, Risk, and Financial Markets." In *The Politics and Science of Prevision: Governing and Probing the Future*, ed. A. Wenger, U. Jasper, and M. D. Cavelty, 141–157. London: Routledge.

Nichols, M. D. (2010). "California's Climate Change Policies: Lessons for the Nation." *Carbon and Climate Law Review: CCLR* **4**(2): 154–160.

Niemeyer, S., and J. Jennstål (2016). "The Deliberative Democratic Inclusion of Future Generations." In *Institutions for Future Generations*, ed. I. González-Ricoy and A. Gosseries, 247–265. Oxford: Oxford University Press.

NikkeiAsia (2014). "IMF Finds More Countries Adopting Managed Floating Exchange Rate System." NikkeiAsia, August 19. Retrieved June 29, 2023, from https://asia.nikkei.com/Business/Markets/Forex/IMF-finds-more-countries-adopting-managed-floating-exchange-rate-system.

Nnadi, N. E., and D. A. Carter (2021). "Climate Change and the Emergence of Fungal Pathogens." *PLoS Pathogens* **17**(4): e1009503.

Nordhaus, W. D. (2017). "Revisiting the Social Cost of Carbon." *Proceedings of the National Academy of Sciences* **114**(7): 1518.

Norwegian Ministry of Finance (2021). *Long-Term Perspectives on the Norwegian Economy 2021*. Oslo: Norwegian Ministry of Finance.

Nowotny, H. (1994). *Time: The Modern and Postmodern Experience*. Cambridge: Polity.

Ogburn, W. F. (1923). *Social Change with Respect to Culture and Original Nature*. New York: B. W. Huebsch.

Ord, T. (2020). *The Precipice: Existential Risk and the Future of Humanity*. London: Bloomsbury.

Oreskes, N., and E. M. Conway (2010). *Merchants of Doubt: How a Handful of Scientists Obscured the Truth on Issues from Tobacco Smoke to Climate Change*. New York: Bloomsbury.

Organisation for Economic Co-operation and Development (OECD) (2023). *Pension Markets in Focus 2022*. Paris: OECD.

Orr, D. W. (2016). *Dangerous Years: Climate Change, the Long Emergency, and the Way Forward*. New Haven: Yale University Press.

Ostrom, E. (2009). "A Polycentric Approach for Coping with Climate Change." World Bank Policy Research Working Paper No. 5095.

Owens, S., and T. Rayner (1999). "'When Knowledge Matters': The Role and Influence of the Royal Commission on Environmental Pollution." *Journal of Environmental Policy & Planning* **1**(1): 7–24.

Oye, K. A. (1986). *Cooperation under Anarchy*. Princeton, NJ: Princeton University Press.
Paterson, M., P. Tobin, and S. D. VanDeveer (2022). "Climate Governance Antagonisms: Policy Stability and Repoliticization." *Global Environmental Politics* 22(2): 1–11.
Patomäki, H. (2006). "Realist Ontology for Futures Studies." *Journal of Critical Realism* 5(1): 1–31.
Pavone, T., and Ø. Stiansen (2022). "The Shadow Effect of Courts: Judicial Review and the Politics of Preemptive Reform." *American Political Science Review* 116(1): 322–336.
Pelling, M., and C. High (2005). "Understanding Adaptation: What Can Social Capital Offer Assessments of Adaptive Capacity?" *Global Environmental Change* 15(4): 308–319.
Peng, W., G. Iyer, V. Bosetti, V. Chaturvedi, J. Edmonds, A. A. Fawcett, S. Hallegatte, D. G. Victor, D. van Vuuren, and J. Weyant (2021). "Climate Policy Models Need to Get Real about People—Here's How." *Nature* 594: 174–176.
Petryna, A. (2022). *Horizon Work: At the Edges of Knowledge in an Age of Runaway Climate Change*. Princeton, NJ: Princeton University Press.
Pierson, P. (2000). "Increasing Returns, Path Dependence, and the Study of Politics." *American Political Science Review* 94(2): 251–267.
Pierson, P. (2004). *Politics in Time: History, Institutions, and Social Analysis*. Princeton, NJ: Princeton University Press.
Pintér, L., M. Kok, and D. Almassy (2017). "Measuring Progress in Achieving the Sustainable Development Goals." In *Governing through Goals: Sustainable Development Goals as Governance Innovation*, ed. N. Kanie and F. Biermann, 99–133. Cambridge, MA: MIT Press.
Porter, T., and L. Stockdale (2016). "The Strategic Manipulation of Transnational Temporalities." *Globalizations* 13(3): 270–284.
Prítyi, M., D. Docherty, and T. Lavery (2021). *Foresight and Anticipatory Governance in Practice: Lessons in Effective Foresight Institutionalisation*. Paris: OECD.
Public Investment Fund (PIF) (2023). "About PIF." Retrieved April 26, 2023, from https://www.pif.gov.sa/en/Pages/AboutPIF.aspx.
Ranchordás, S. (2015). "Sunset Clauses and Experimental Regulations: Blessing or Curse for Legal Certainty?" *Statute Law Review* 36(1): 28–45.
Rees, M. (2003). *Our Final Century? Will the Human Race Survive the Twenty-First Century?* London: William Heinemann.
Rocco, P. (2021). "Keeping Score: The Congressional Budget Office and the Politics of Institutional Durability." *Polity* 53(4): 691–717.
Rockström, J., W. Steffen, K. Noone, Å. Persson, F. S. Chapin, E. F. Lambin, T. M. Lenton, et al. (2009). "A Safe Operating Space for Humanity." *Nature* 461(7263): 472–475.
Roger, C., T. Hale, and L. B. Andonova (2017). "The Comparative Politics of Transnational Climate Governance." *International Interactions* 43(1): 1–25.
Romelli, D. (2022). "The Political Economy of Reforms in Central Bank Design: Evidence from a New Dataset." Paper presented at the 74th Economic Policy Panel Meeting, October 21 and 22.
Rosa, H. (2013). *Social Acceleration: A New Theory of Modernity*. New York: Columbia University Press.
Roser, M. (2021). "Economic Growth." Retrieved April 2, 2021, from https://ourworldindata.org/economic-growth.

Rousell, D., and A. Cutter-Mackenzie-Knowles (2020). "A Systematic Review of Climate Change Education: Giving Children and Young People a 'Voice' and a 'Hand' in Redressing Climate Change." *Children's Geographies* **18**(2): 191–208.

Royal Dutch Shell (2021). *The Energy Transformation Scenarios.* London: Royal Dutch Shell. Retrieved August 6, 2023, from https://www.shell.com/energy-and-innovation/the-energy-future/scenarios/what-are-the-previous-shell-scenarios/the-energy-transformation-scenarios.html#vanity-aHR0cHM6Ly93d3cuc2hlbGwuY29tL2VuZXJneS1hbmQtaW5ub3ZhdGlvbi90aGUtZW5lcmd5LWZ1dHVyZS9zY2VuYXJpb3MvdGhlLWVuZXJneS10cmFuc2Zvcm1hdGlvbi1zY2VuYXJpb3MuaHRtbA=&iframe=L3dlYmFwcHMvU2NlbmFyaW9zX3B2bmdfG9yaXpvbnMv.

Ruggie, J. (1993). *Multilateralism Matters: The Theory and Praxis of an Institutional Form.* New York: Columbia University Press.

Sabel, C. F., and W. H. Simon (2011). "Minimalism and Experimentalism in the Administrative State." *Georgetown Law Journal* **100**(1): 53–94.

Sabel, C. F., and D. G. Victor (2017). "Governing Global Problems under Uncertainty: Making Bottom-Up Climate Policy Work." *Climatic Change* **144**(1): 15–27.

Sabel, C. F., and D. G. Victor (2022). *Fixing the Climate: Strategies for an Uncertain World.* Princeton, NJ: Princeton University Press.

Sabel, C. F., and J. Zeitlin (2012). "Experimentalist Governance." In *The Oxford Handbook of Governance,* ed. D. Levi-Faur, 169–184. Oxford: Oxford University Press.

Sachs, J. D., S. S. A. Karim, L. Aknin, J. Allen, K. Brosbøl, F. Colombo, G. C. Barron, et al. (2022). "The *Lancet* Commission on Lessons for the Future from the COVID-19 Pandemic." *Lancet* **400**(10359): P1224–1280.

Sælen, H., J. Hovi, D. Sprinz, and A. Underdal (2020). "How US Withdrawal Might Influence Cooperation under the Paris Climate Agreement." *Environmental Science & Policy* **108**: 121–132.

Sampson, R. C., and Y. Shi (2023). "Are U.S. Firms Becoming More Short-Term Oriented? Evidence of Shifting Firm Time Horizons from Implied Discount Rates, 1980–2013." *Strategic Management Journal* **44**: 231–263.

Sandler, T. (2004). *Global Collective Action.* Cambridge: Cambridge University Press.

Sawin, J. L. (2001). "The Role of Government in the Development and Diffusion of Renewable Energy Technologies: Wind Power in the United States, California, Denmark and Germany, 1970–2000." PhD diss., Tufts University.

Schedler, A., and J. Santiso (1998). "Democracy and Time: An Invitation." *International Political Science Review* **19**(1): 5–18.

Schipper, E. L. F. (2006). "Conceptual History of Adaptation in the UNFCCC Process." *Review of European Community & International Environmental Law* **15**(1): 82–92.

School of International Futures (2022). *Fair Public Policies for All Generations: An Assessment Framework.* London: School of International Futures and Calouste Gulbenkian Foundation.

Sebenius, J. K. (1991). "Designing Negotiations toward a New Regime: The Case of Global Warming." *International Security* **15**(4): 110–148.

Setzer, J., and L. Benjamin (2020). "Climate Litigation in the Global South: Constraints and Innovations." *Transnational Environmental Law* **9**(1): 77–101.

Setzer, J., and C. Higham (2021). *Global Trends in Climate Change Litigation: 2021 Snapshot.* London: Grantham Research Institute on Climate Change and the Environment, London School of Economics and Political Science.

Shoham, S., and N. Lamay (2006). "Commission for Future Generations in the Knesset: Lessons Learnt." In *Handbook of Intergenerational Justice,* ed. J. Tremmel, 244–281. Cheltenham, UK: Edward Elgar.

Shue, H. (2014). *Climate Justice: Vulnerability and Protection.* Oxford: Oxford University Press.

Simmons, B., and Z. Elkins (2004). "The Globalization of Liberalization: Policy Diffusion in the International Political Economy." *American Political Science Review* 98(1): 171–189.

Simmons, J. W. (2016). *The Politics of Technological Progress: Parties, Time Horizons and Long-Term Economic Development.* Cambridge: Cambridge University Press.

Skocpol, T. (1996). *Boomerang: Clinton's Health Security Effort and the Turn against Government in US Politics.* New York: Norton.

Slaughter, A.-M. (2017). *The Chessboard and the Web: Strategies of Connection in a Dangerous World.* New Haven: Yale University Press.

Smith, G. (2021). *Can Democracy Safeguard the Future?* Cambridge: Polity.

Smith, S. M., O. Geden, G. F. Nemet, M. J. Gidden, W. F. Lamb, C. Powis, R. Bellamy, et al. (2023). *The State of Carbon Dioxide Removal, 1st ed.* N.p.: The State of Carbon Dioxide Removal. Retrieved June 27, 2023, from https://www.stateofcdr.org/.

Steffen, W., K. Richardson, J. Rockström, S. E. Cornell, I. Fetzer, E. M. Bennett, R. Biggs, et al. (2015). "Planetary Boundaries: Guiding Human Development on a Changing Planet." *Science* 347(6223): 1259855.

Steffen, W., A. Sanderson, P. D. Tyson, J. Jäger, P. A. Matson, B. Moore III, F. Oldfield, et al. (2004). *Global Change and the Earth System: A Planet under Pressure.* New York: Springer.

Steven, D., and B. Francruz (2021). *Future Thinking and Future Generations: Towards a Global Agenda to Understand, Act for, and Represent Future Generations in the Multilateral System.* New York: United Nations Foundation.

Stockdale, L. (2016). *Taming an Uncertain Future: Temporality, Sovereignty, and the Politics of Anticipatory Governance.* London: Rowman & Littlefield.

Stokes, L. C. (2020). *Short Circuiting Policy: Interest Groups and the Battle over Clean Energy and Climate Policy in the American States.* Oxford: Oxford University Press.

Szabó, M. (2016). "A Common Heritage Fund for Future Generations." In *Institutions for Future Generations,* ed. I. González-Ricoy and A. Gosseries, 197–213. Oxford: Oxford University Press.

Tarrow, S. (1998). *Power in Movement: Social Movements and Contentious Politics.* Cambridge: Cambridge University Press.

Tarsney, C. (2022). "The Epistemic Challenge to Longtermism." Global Priorities Institute Working Paper No. 3-2022.

Tetlock, P. E. (2005). *Expert Political Judgment.* Princeton, NJ: Princeton University Press.

Thomas, D. C. (2001). *The Helsinki Effect: International Norms, Human Rights, and the Demise of Communism.* Princeton, NJ: Princeton University Press.

Thomas, L. (2019). "Stop Panicking about Corporate Short-Termism." *Harvard Business Review,* June 28. Retrieved June 27, 2023, from https://hbr.org/2019/06/stop-panicking-about-corporate-short-termism.

Toffler, A. (1970). *Future Shock*. New York: Bantam.
Tonn, B. (1996). "A Design for Future-Oriented Government." *Futures* **28**(5): 413–431.
Tremmel, J. (2006). "Establishing Intergenerational Justice in National Constitutions." *Handbook of Intergenerational Justice*, ed. J. Tremmel, 187–214. Cheltenham, UK: Edward Elgar.
Tremmel, J. (2017). "Constitutions as Intergenerational Contracts: Flexible or Fixed?" *Intergenerational Justice Review* **3**(1): 4–17.
Tremmel, J. (2018). "Updating Democracy for Future Generations: Adding a Fourth Branch to the Separation of Powers Model." *Items: Insights from the Social Sciences*, February 6. Retrieved June 27, 2023, from https://items.ssrc.org/democracy-papers/updating-democracy-for-future-generations-adding-a-fourth-branch-to-the-separation-of-powers-model/.
Tsebelis, G. (2002). *Veto Players: How Political Institutions Work*. Princeton, NJ: Princeton University Press.
Tverberg, A. (2017). *The Use of Sunset and Evaluation Clauses in the Norwegian Legal System*. Seoul: Korean Legislative Research Institute.
United Nations (2015). *The Millennium Development Goals Report*. New York: United Nations.
United Nations (2022a). *Our Common Agenda*. New York: United Nations.
United Nations (2022b). *The Sustainable Development Goals Report*. New York: United Nations.
United Nations Framework Convention on Climate Change (UNFCCC) (2019). *Yearbook of Climate Action 2019*. Bonn: UNFCCC.
United Nations Secretary-General (2013). *Intergenerational Solidarity and the Needs of Future Generations*. New York: United Nations General Assembly.
Urpelainen, J. (2009). "Explaining the Schwarzenegger Phenomenon: Local Frontrunners in Climate Policy." *Global Environmental Politics* **9**(3): 82–105.
Urpelainen, J. (2013). "A Model of Dynamic Climate Governance: Dream Big, Win Small." *International Environmental Agreements* **13**: 107–125.
Valenzuela, J. M. (2020). "Regulatory Capacity and Knowledge Brokers in the Decarbonization of Electricity Systems." PhD diss., Blavatnik School of Government, University of Oxford.
van Asselt, H. (2016). "The Role of Non-state Actors in Reviewing Ambition, Implementation, and Compliance under the Paris Agreement." *Climate Law* **6**(1): 91–108.
van Beek, L., M. Hajer, P. Pelzer, D. van Vuuren, and C. Cassen (2020). "Anticipating Futures through Models: The Rise of Integrated Assessment Modelling in the Climate Science-Policy Interface since 1970." *Global Environmental Change* **65**: 102191.
van der Ven, H., S. Bernstein, and M. Hoffmann (2017). "Valuing the Contributions of Nonstate and Subnational Actors to Climate Governance." *Global Environmental Politics* **17**(1): 1–20.
Venkataraman, B. (2020). *The Optimist's Telescope: Thinking ahead in a Reckless Age*. New York: Riverhead Books.
Verbeeten, F. H. M., and R. F. Speklé (2015). "Management Control, Results-Oriented Culture and Public Sector Performance: Empirical Evidence on New Public Management." *Organization Studies* **36**(7): 953–978.
Vickers, G. (1970). *Value Systems and Social Processes*. London: Pelican.
Victor, D. G. (2011). *Global Warming Gridlock: Creating More Effective Strategies for Protecting the Planet*. Cambridge: Cambridge University Press.

Victor, D. G. (2015). "Why Paris Worked: A Different Approach to Climate Diplomacy." *Yale E360*, December 15. Retrieved June 27, 2023, from https://e360.yale.edu/features/why_paris_worked_a_different_approach_to_climate_diplomacy.

Victor, D. G., F. W. Geels, and S. Sharpe (2019). *Accelerating the Low Carbon Transition: The Case for Stronger, More Targeted and Coordinated International Action*. London: Energy Transitions Commission

Victor, D. G., K. Raustiala, and E. B. Skolnikoff (1998). *The Implementation and Effectiveness of International Environmental Commitments*. Cambridge: MIT Press.

Vogel, D. (2018). *California Greenin'*. Princeton, NJ: Princeton University Press.

Voß, J. P., D. Bauknecht, and R. Kemp (2006). *Reflexive Governance for Sustainable Development*. Cheltenham, UK: Edward Elgar.

Warde, P., and S. Sörlin (2015). "Expertise for the Future: The Emergence of Environmental Prediction c. 1920–1970." In *The Struggle for the Long-Term in Transnational Science and Politics: Forging the Future*, ed. J. Andersson and E. Rindzevičiūtė, 38–62. London: Routledge.

Weaver, K. (1989). "Setting and Firing Policy Triggers." *Journal of Public Policy* **9**(3): 307–336.

Wegrich, K. (2019). "Incrementalism and Its Alternatives." In *The Oxford Handbook of Time and Politics*, ed. K. H. Goetz. Oxford: Oxford University Press.

Wells, H. G. (1902). *The Discovery of the Future*. London: T. Fisher Unwin.

Wells, R. (2022). "Citizens' Assemblies and Juries on Climate Change: Lessons from Their Use in Practice." In *Addressing the Climate Crisis: Local Action in Theory and Practice*, ed. C. Howarth, M. Lane, and A. Slevin, 119–128. Cham: Springer.

Wells, R., C. Howarth, and L. I. Brand-Correa (2021). "Are Citizen Juries and Assemblies on Climate Change Driving Democratic Climate Policymaking? An Exploration of Two Case Studies in the UK." *Climatic Change* **168**(1): 5.

Wenger, A., U. Jasper, and M. Dunn Cavelty (2020). *The Politics and Science of Prevision: Governing and Probing the Future*. London: Routledge.

Westminster Foundation for Democracy. (2021). *An Introduction to Deliberative Democracy for Members of Parliament*. London: Westminster Foundation for Democracy.

Willis, R., N. Curato, and G. Smith (2022). "Deliberative Democracy and the Climate Crisis." *WIREs Climate Change* **13**(2): e759.

Wong, R. B., P.-É. Will, J. Lee, J. Oi, and P. Perdue (1991). "Chinese Traditions of Grain Storage." In *Nourish the People: The State Civilian Granary System in China, 1650–1850*, 1–16. Ann Arbor: University of Michigan Press.

Wood, C. (2002). *Environmental Impact Assessment: A Comparative Review*. New York: Pearson Education.

Woods, N. (2006). *The Globalizers: The IMF, the World Bank, and Their Borrowers*. Ithaca: Cornell University Press.

Wu, J., C. Zuidema, K. Gugerell, and G. de Roo (2017). "Mind the Gap! Barriers and Implementation Deficiencies of Energy Policies at the local Scale in Urban China." *Energy Policy* **106**: 201–211.

WWF (2020). *The Living Planet Report 2020, Living Planet Index*. Gland, Switzerland: WWF.

Young, K. E. (2020). "Sovereign Risk: Gulf Sovereign Wealth Funds as Engines of Growth and Political Resource." *British Journal of Middle Eastern Studies* **47**(1): 96–116.

Young, O. R. (2017). "Conceptualization: Goal Setting as a Strategy for Earth System Governance." In *Governing through Goals: Sustainable Development Goals as Governance Innovation*, ed. N. Kanie and F. Biermann, 31–51. Cambridge, MA: MIT Press.

Young, O. R., and F. Biermann (2017). *Governing Complex Systems: Social Capital for the Anthropocene*. Cambridge, MA: MIT Press.

Young, O. R., A. Underdal, N. Kanie, and R. E. Kim (2017). "Goal Setting in the Anthropocene: The Ultimate Challenge of Planetary Stewardship." In *Governing through Goals: Sustainable Development Goals as Governance Innovation*, ed. N. Kanie and F. Biermann, 53–74. Cambridge, MA: MIT Press.

Zenghelis, D., R. Fouquet, and R. Hippe (2018). "Stranded Assets: Then and Now." In *Stranded Assets and the Environment: Risk, Resilience and Opportunity*, ed. B. Caldecott, 23–54. London: Taylor & Francis.

Zhao, X., O. R. Young, Y. Qi, and D. Guttman (2020). "Back to the Future: Can Chinese Doubling Down and American Muddling Through Fulfill 21st Century Needs for Environmental Governance?" *Environmental Policy and Governance* **30**(2): 59–70.

Zhou, C., and Z. Wang (2020). "China Suffers Worst Power Blackouts in a Decade, on Post-coronavirus Export Boom, Coal Supply Shortage." *South China Morning Post*, December 23. Retrieved June 27, 2023, from https://www.scmp.com/economy/china-economy/article/3115119/china-suffers-worst-power-blackouts-decade-post-coronavirus.

INDEX

Abbott, Kenneth, 67
acceleration, 145, 195–98
activism, political, 99–100
Adam, Barbara, 158, 165
adaptation: climate change and, 22–23, 136–37, 154; determinants of, 148; durability and, 122–27, 176–77; endurance and, 105; to global temperatures, 144–45; investments in, 42; plans for, 173; rates of change and, 147–49
adaptive management, 55
Agenda 2063 (Africa), 107
agenda-setting, 27
agent-based modeling, 141, 161, 162
Aklin, Michaël, 132
Alien Tort Claims Act, 88
American Constitution, 13
analysis: of the future, 157–65; hubris in, 160; seriousness of, 129–37; of war, 129–30
Annex I countries, 37
Anthropocene, start of, 185–86
anticipatory governance, 43–44. *See also* governance
antimicrobial resistance, 146
Asimov, Isaac, 157
asset revaluation model, 134–35, 136, 153
Auld, Graeme, 133
autocorrection, 46

bacteria, 188
Bangladesh, 108
Barrett, Scott, 163
Bayesian approach, 161

Beck, Ulrich, 190–91, 200n29
belief systems, 14
benchmarking, 68, 111
benefit corporations (B corps), 92
Bernstein, Steven, 62, 63, 133, 161, 164
Better Business Act (United Kingdom), 92
biodiversity, 185
bioenergy with carbon capture and storage (BECCS), 42, 154
Bjornerud, Marcia, 128, 186, 187
boomerang effect, 72
Boston, Jonathan, 23, 52, 76–77, 191
Burke, Edmund, 70
Busby, Joshua, 133

California Air Resources Board (CARB), 84–86
campaign contributions, 175
capacity-building, 95
capital, reserves of, 124–25
capitalism, 1, 92
carbon, 1, 2, 49, 126–27
Carney, Mark, 23
Cashore, Benjamin, 133
casual mechanisms, 140, 164
catalysts: for climate change, 63–64; in early action paradox, 167; eroding of obstructionism by, 60–69; flexibility regarding, 65; as institutionalized, 64; iteration of, 66; learning effects from, 62–63; movers in, 65; overview of, 60–61; policy interventions and, 62
catalytic cooperation model, 133–34, 136

233

central banks, 82, 83, 86, 172
Central Organization Department (China), 109–10
Charter of the United Nations, 13
checks and balances, 88
Chenoweth, Erica, 99
Chile, wealth funds in, 124
China: accomplishments of, 107; authoritarian nature of, 118–19; Central Organization Department in, 109–10; competing demands in, 114; COVID-19 pandemic and, 114; emissions trading system of, 63; five-year plan of, 107; fragmented authoritarianism of, 106; governance in, 13; government structure in, 26; net-zero target of, 109; organizational patterns in, 56; Paris Agreement and, 109; planning process in, 106; policymaking of, 106, 118–19; wealth funds in, 124
Citizen's Assembly on Climate (Germany), 96
civic engagement, 99–100
Cixin, Liu, 144
clauses, review and sunset, 115–18
Clean Air Act (United States), 80
Climate Assembly (Denmark), 97
Climate Assembly UK, 97
climate change: activism regarding, 99–100; adaptation and, 22–23, 136–37, 154; analysis of, 131–37; asset threats and, 133–34; as catalyst for institutional change, 177–80; catalytic action regarding, 63–64; catalytic cooperation model regarding, 133–34; consequences of, 2; defining, 9; deliberative techniques regarding, 97–99; disclosure requirements regarding, 93; distributional view of, 133–34; dynamic problem structure chains of, 151–56; education regarding, 95–97; elements of, 6; as entertainment subject, 143; experimentalism for, 57–58; foresight and, 50; as global collective action problem, 131–32; governance and, 3, 168–69; high-mitigation scenario regarding, 153–54; incentives regarding, 110; institutional agenda on, 169–77; institutional lag and, 36; integrated assessment models (IAMs) for, 141–42; knowledge base regarding, 45; levels of, 146; litigation regarding, 87–88; low-mitigation scenario regarding, 155–56; mitigation of, 62, 132, 133, 134, 147–49, 154; norm cascade for, 63; object of policy in, 61; policy winners and losers regarding, 132–33; polycentric model of, 132; prevention and, 152–53; rates of change and, 146, 147, 149; salience from, 46; shadow interests and, 31; Special Report on Global Warming of 1.5°C (IPCC) and, 72–74; strategies regarding, 178–79; success regarding, 180–81; as "super wicked" problem, 61, 134, 200n26; targets for, 63; as transition problem, 4; transnational governance of, 65; triggers and reserves regarding, 126–27; trusteeship and, 86; understanding regarding, 7–9
Climate Change Act, 74
Coase, Ronald, 31
coercion, in experimentalist governance, 56
Cold War, 120, 144, 189
Colgan, Jeff, 133
collective action, 131–32, 133–34, 153
Collect Pond, Manhattan, New York, 39
Collinridge, David, 41
Colombia, litigation in, 87
Commission for Sustainable Development and the Rights of Future Generations (Tunisia), 75
commitment devices, 80
commitment-making, iteration of, 66
Committee for the Future (Finland), 50–51, 75
common pool resources, 131
communication, changes in, 1–2
companies/institutions/organizations: accountability cycles of, 24–25, 26–27; building strategies of, 104; campaign contributions and, 175; catalyst overview of, 68–69; changing of, 35, 171–72; climate goal mandates for, 175;

commitment-making by, 66; constituency growth by, 68; earnings reporting of, 91–92; endurance of, 105–14; foresight use by, 50; goal-setting and benchmarking by, 68; inconsistent incentives for, 27; issue processing by, 27; malleability of, 120; norm changes by, 95; path dependence of, 35; political weight of, 77; and power, 34; and processes for learning, 67; punctuated equilibrium and, 35–36; reflexive governance and, 115–22; reporting by, 26–27; short-termism and, 24–28; as sticky, 34–35, 36; thermostatic, 122; updating of, 115–22
computational models, 161
Conference of the Parties (COP), 65, 67, 73, 97
Congressional Budget Office (CBO), 47–48, 72
conservatism, 14
constitutions, change process for, 35
constructivism, limitations of, 7
COP. *See* Conference of the Parties (COP)
Council on Future (Sweden), 75
COVID-19 pandemic, 46, 47, 112, 114, 122–23
creeping problems, 200n26
crises, salience from, 46
critical junctures, 139, 140
critical theory, 16

Dahl, Robert, 169
Dannenberg, Astrid, 163
debt, during COVID-19 pandemic, 47
decarbonization, 57–58, 62, 153, 154, 178
deceleration, rates of change and, 145
decision-making, 25–26, 45–46, 47, 79, 83
Declaration of Independence, 193
Declaration of the Rights of Man and of the Citizen, 193
Declaration on Future Generations, 15, 89
delegation, 79, 80, 84
deliberative techniques, 97–99
Delphi method, 49
democracy, 130, 196

democratic myopia thesis, 25
Denmark, Climate Assembly of, 97
Dewey, John, 55
diffusion, in experimentalism, 58
dilemma of control, 41
direct air capture (DAC), 154
discount rates, 138
distributional politics, 130–31, 133–34, 136, 153
Douglass, Frederick, 30
drought, 9, 123, 178
Dubash, Navroz, 174
durability, 122–27, 176–77
dynamic problem structure chains, 149–65

early action paradox, 3, 22, 38, 43; and anticipatory governance, 43–44; and catalysts, 61; and experimentalism, 57; and informational instruments, 52; solutions to, 47, 61, 167; and uncertainty, 54
Earth Overshoot Day, 184
earth system governance, 186
ecologism, 14
economic inequality, 131
economy, 125, 145, 150–51, 188
education, for climate change, 95–97
Edwards, Paul, 187–88
Egypt 2030 plan, 107
elections, democratic myopia thesis regarding, 25
electricity, decarbonization of, 39
emissions: acceleration of, 146; from Annex I countries, 37; as climate change cause, 9; doubling of, 21; economics and, 23; reduction of, 8, 39; statistics regarding, 39, 185; technologies regarding, 42; trading systems for, 63; triggers regarding, 126–27. *See also* climate change
endurance, institutional, 105–14
environmental change, 190
environmental impact assessments (EIAs), 48, 52, 53
environmental movement, 184
environmental policy, 118

Environmental Protection Agency (EPA), 80
epidemiological models, 161
Equiano, Olaudah, 30
ethanol, 176
European Court of Human Rights, 84
European Union (EU), Green Deal of, 170
experimentalism: agenda for, 174; coercion in, 56; defined, 153; in early action paradox, 167; failures of, 41, 176; goal in, 56; learning and diffusion in, 58; overview of, 54–60; political consensus in, 58–59; problem structure in, 59; willingness for, 57
experimental physics, 160
extinction, 185
Extinction Rebellion, 100

Fawcett, Paul, 196
Figueres, Christiana, 94–95
financial crisis (2008–2009), 46
Finland, 50–51, 52, 75
Five Points slum, Manhattan, New York, 39
food production, 145
food storage, 123
forecasting, 49
foresight, 44–54, 167, 172, 203n30
forest fires, 10, 187
formal theory, 141
free riding, 135
French Citizens Convention on Climate, 96
Fridays for Future, 100
Fuerth, Leon, 44
future/future generations: arguments regarding, 53; caring about, 14–15, 192; empirical analysis of, 157–65; failures regarding, 176; hubris regarding, 160; innovation regarding, 160; as known and salient, 44–54; outcomes for, 158–59, 160; powerlessness of, 29; pragmatic approach to, 159; predictions regarding, 157–58; quantitative methods regarding, 157; representation of, 71–78; selective interests regarding, 53–54; study of, 159; voice for, 71–78

game theoretic modeling, 141
gaming, 142–43
Geels, Frank, 133
geoengineering, 179
geopolitical rivalry, 144
Germany, 87, 96
Glasgow Financial Alliance for Net Zero, 111
Global Footprint Network, 184
globalization, 2, 55
Global Optimism, 94–95
Global Stocktake (Paris Agreement), 67, 169
global temperatures, 9, 144–45, 146, 155, 178, 185. *See also* climate change
goals/goal-setting: behavior of actors and, 109–10; benchmarking and, 111; benefits of, 173; coherence regarding, 111–12; competing, 114; defined, 56; endurance of, 106, 108; events for, 107; in experimentalist governance, 56; incentives regarding, 110; institutionalization of, 106; institutional lag and, 105–14, 167; by institutions, 68; overview of, 64, 106; as problematic, 114; rule-making versus, 106; salience of, 108; targets for, 173
Goetz, Klaus, 197
Gore, Al, 24
governance: adaptive management in, 55; anticipatory, 43–44; challenges of, 3; experimentalism in, 54–60; future as known and salient in, 44–54; intergovernmental, 3; of long problems, 12–15; participatory system of, 55; procedural requirements in, 47; reflexive, 55, 115–22, 167; of risk society, 190–91; self-constraint in, 81; support for, 25; temporal inconsistency of preferences of, 27; time-inconsistency problem in, 81; transnational, 3, 65; use of intermediaries in, 65
Government Foresight Group (Finland), 51
Great Acceleration, 185
Green, Jessica, 133
Green Climate Fund, 67
Green Deal (EU), 170

greenhouse gases (GHGs), 6, 154–55.
  See also carbon; emissions
Green New Deal (United States), 170
Gross Domestic Product (GDP), 47, 185
Guardian of Future Generations (Malta), 75

Held, David, 192
Helimann, Sebastian, 56
Hellmann, Gunther, 159
Hitler, Adolf, 13
Hoffmann, Matthew, 62, 63, 133
Holocene, 147, 185
horse manure, 51
hubris, in analysis, 160
human development, pace and scale of, 185
human rights institutions, 86
humans, rights and timescales of, 184–87, 193
Hungary, 75, 76
Huntington, Samuel, 196

incentives, 110, 131, 133, 179
industrialization, 9, 196
inequality, 4, 6, 130–31
inflation, 83
innovation, in social science methods, 160
institutional change, climate change as catalyst for, 177–80
institutional lag, 4, 32–37, 102–5, 167; climate change and, 36; endurance and, 105–14; example of, 22; goal-setting and, 105–14, 167; reflexive governance and, 115–22, 167; triggers and reserves of, 122–27, 167; updating institutions and, 115–22
institutions. *See* companies/institutions/organizations
integrated assessment models (IAMs), 141–42, 161–62
Inter-American Court of Human Rights, 84
Intergovernmental Panel on Climate Change (IPCC), 21, 72–74, 77–78, 94, 188
International Energy Association, 146
International Monetary Fund (IMF), 83, 121
international relations theory, 33–34
intervention points, sensitive, 61

Irish Citizens' Assembly (Ireland), 97, 99
Iroquois people, maxim of, 14, 47
Ise Jinju temple (Japan), 102
Israel, 75, 76

Jacobs, Alan, 7, 128
Jacobs, Harriet, 30
Japan, in World War II, 143
Jasanoff, Sheila, 200n29
Jefferson, Thomas, 102–3, 115
Johnson, Steven, 39
Jordan, 123
Juliana v. the United States, 87

Kant, Immanuel, 130
Kennedy, John, 108
Keohane, Nannerl O., 94
Knesset Commission for Future Generations (Israel), 75, 76
Korea, wealth funds in, 124
Krznaric, Roman, 28, 32, 192
Kyoto Protocol, 37, 57, 59, 67, 119, 135

land restoration, 58
land use policies, 58
Lasswell, Harold, 128
laws, change process for, 35
legal long-termism, 88–90
Levin, Kelly, 133
Lindblom, Charles, 38–39, 43
literacy, 139
lock-in, overcoming, 133
long problems: cause and effect related to, 9; costs and benefits of, 12; early action and, 38; existing tools for, 137–43; governance challenges regarding, 3; governance need regarding, 12–15; governance process regarding, 4; as marathons, 143; ongoing problems as compared to, 10; political dilemmas regarding, 22; problem structure shifting of, 33; study of, 4–5
loss and damage, 8, 179
Lowi, Theodore, 116

Maastricht Treaty, 80
Madison, James, 102–3, 115, 120–21
malapportionment, 175
malleability, 120
Malta, 75
Malthus, Robert Thomas, 145, 150, 197–98
Manulak, Michael, 117
Mao Zedong, 13
Marrakech Partnership for Global Climate Action (UNFCCC), 60, 65
Marx, Karl, 179–80
material transfers, process of, 66–67
media, influence of, 28
meteorology, 187–88
microorganisms, 188
Middle Ages, pathogen spread in, 195
Mildenberger, Matto, 132
Milieudefensie et al. v. Royal Dutch Shell plc., 87
Millennium Development Goals (MDGs), 105–6, 107, 112, 113
Ministry of Cabinet Affairs and the Future (United Arab Emirates), 75
mitigation: of climate change, 22–23, 132, 133, 134, 147–49, 154; dynamics of, 31; plans for, 173; viewpoints regarding, 8
mobilization, 99
Montreal, Canada, 56–57
Montreal Method, 57
Montreal Protocol, 56–57, 135, 152
moral cosmopolitanism, 193
moral long-termism, 194
muddling through, 37–42, 43
multilateral rule-making, 119

National Adaptation Plan (Bangladesh), 108
national adaptation plans, 108
National Environmental Policy Act, 48
National Foresight Network (Finland), 51
nationally determined contribution (NDC) structure, 59
natural disasters, 123
natural resources, human budget of, 184
near-term decisions, future implications of, 44–54

New York City, horse manure problem in, 51
New Zealand, 48–49, 52, 72, 114
Nigeria, 125
NIMBYism, 117
nonsimultaneous exchange, 31
North, Douglas, 128
North Sea, 23
Norway, 124, 125, 126
Nowotny, Helga, 186
nuclear age, 189

obstructionism, 37–42, 41, 60–69, 174
ocean chemistry, 185
Ogburn, William Fielding, 196, 197, 198
ongoing problems, long problems as compared to, 10
Ord, Toby, 193
Organisation for Economic Co-operation and Development (OECD), 109, 124, 203n30
Organization of Africa Unity, 107
organizations. *See* companies/institutions/organizations
overlapping generations, 138

pandemic disease, 6, 40, 188, 195. *See also* COVID-19 pandemic
Paris Agreement: as abstract, 60; activism and, 99–100; adoption of, 110; benchmarking for, 111; capacity-building and, 95; catalyst overview of, 69; commitment-making and, 66; effects of, 37; flexibility regarding, 65; Global Stocktake of, 67, 169; as institutionalized, 64; integrated assessment models (IAMs) for, 142; iteration of, 66; legitimacy of, 64; as modest instrument, 59–60; overview of, 108; Paris Committee on Capacity Building, 67; review of, 119; United States and, 111–12
Paris Committee on Capacity Building, 67, 95
Parliamentary Commissioner for Future Generations (Hungary), 75, 76
participatory deliberation, 172

INDEX 239

path dependence, 139, 140
Patomäki, Heikki, 161
Patrocínio, José do, 30
Pearl Harbor attack, 143
penalty defaults, 56, 59
pension funds, 126, 137
Permanent School Fund (Texas), 124
Permanent University Fund (Texas), 124
petroleum, 123
phosphorous cycle, 185
Pierson, Paul, 10, 12, 128, 140
Pinochet, Augusto, 88
planetary systems, 13, 184–87
planning, acceleration and, 196
Plato, 116
plotlines, 164
Polanyi, Karl, 179–80
policy: acceleration and, 196; barriers to, 105; environmental, 118; reserves in, 122–27; triggers in, 122–27. *See also specific policies*
political activism, 99–100
political consensus, in experimentalism, 58–59
political leaders, tenure of, 108–9
political problems, 5–12, 33–34, 191–94, 195–98
political will, 170
politics: capacity for change in, 104; carbon view of, 1–2; defined, 29; methods for change in, 32; veto players in, 104
Pompeii, 44–45
population growth, 145, 150–51
poverty, 112
power relationships, 34, 175–76
prediction, evaluation of, 157–58
preferences, 24, 90–101, 138
prevention, of climate change, 152–53
primacy, 9
Prince, Mary, 30
problem length, 3, 5, 9, 10, 11, 38, 129–39
problems, defining, 5–12. *See also* political problems
problem-solving, participatory system of, 55

Public Investment Fund (PIF) (Saudi Arabia), 125, 126
punctuated equilibrium, 35–36
pure time preferences, 24

qualitative process tracing, 140

Rapa Nui (Easter Island), 185
rapid change, 196
rates of change, 143–49
rationality, bounded, 24
Rayner, Steve, 136
reflexive governance, 55, 115–22, 167
regulatory impact assessments (RIAs), 48–49
renewable energy, 41–42, 61, 62, 67, 146, 148
reserves, 122–27, 167, 174
resistance development, 144–45
review clauses, 115–18, 173–74
risk, future, 140
risk society, 190–91
Rivett-Carnac, Tom, 94–95
Roosevelt, Franklin, 82–83
Rosa, Hartmut, 195, 198
Ruggie, John, 122
rule-making, goal-setting versus, 106
Russia, wealth funds in, 124

Sabel, Charles, 54, 56, 136
salience, 37–42, 44–54, 108
SARS, 46
Saudi Arabia, 107, 125, 126
scenario planning, 142
scenario-style approach, 163–64
School of International Futures, 210n3
science, future vision of, 187–91
Science Based Targets Initiative, 110
scientific development, 13–14, 43–44
seawalls, 58
securities law, 91–92
security, foresight regarding, 50
self-control, shadow interests and, 81
sensitive intervention points, 61

shadow interests: activism regarding, 100; changing preferences regarding, 90–101; example of, 22; horizon-shifting regarding, 90–101; interventions regarding, 70–71; overview of, 4, 29–32, 70–71, 167; political focus of, 70; representation and, 71–78, 167; self-control strategies and, 81; time horizons in, 167; trusteeship and, 78–90, 167; voices for future generations regarding, 71–78
shadow of future, 191–94
shareholder capitalism, 92
short-termism: causes of, 138; in climate, 23; crises as affecting, 46; defined, 3; drivers behind, 21; incentives for, 179; limitations of, 28–29; overcoming, 91; tyranny of the present and, 22–29; variation in, 25
simulation modeling, 141, 142–43, 162–63
Skocpol, Theda, 48
slavery, 29–30, 136
slow-moving casual processes, 140
social media, 190, 195
social science: agent-based models for, 161, 162; analysis and, 129–37; approaches to, 160–61, 164–65; bias in, 129; computational models for, 161; dynamic problem structure chains and, 149–65; empirical analysis of the future in, 157–65; epidemiological models for, 161; existing tools in, 137–43; free riding and, 135; innovation in methods of, 160; overview of, 128; rates of change and, 143–49; scenario-style approaches in, 163–64; simulation modeling for, 162–63; time horizons and, 138–39; timing, sequence, and contingency tools in, 139
social security (United States), 125–26
social systems, timeframe for, 186
social trust, 121–22
society, 1, 70
sociopolitical tipping point, 148
Solyndra, 41
South Africa, deliberative techniques of, 97
Soviet Union, 120, 144

Spain, net-zero transition plan requirement in, 93
Special Report on Global Warming of 1.5°C (IPCC), 72–74
spending, assessment of, 47
stakeholder capitalism, 92
state capacity, as enabling condition, 177
Stockdale, Liam, 53, 140
storytelling, 142–43
Structural Reform Support Division (OECD), 203n30
structure change, of long problems, 149–65
successive limited comparisons, 38–39
Sunrise Movement, 100
sunset clauses, 115–18
support, in governance, 25
sustainable development, 194
Sustainable Development Goals (SDGs), 64, 107, 112–13
Sweden, 75
Syria, drought in, 9

Talanoa Dialogue, 97
taxation, 92
Technical Expert Meetings (UNFCCC), 60
technology/technological development: accumulative infrastructure of, 187; anticipatory governance regarding, 43–44; climate change and, 187–88; effects of, 13–14, 189–90; future vision of, 187–91; learning curve of, 62; network effects of, 62; obstructionism and, 41; rate of change in, 144; salience and, 41; uncertainty and, 41
temporal focal points, 117–18, 139
tenure of statutes act, 116
Tetlock, Philip, 49, 157, 159, 160
Thailand, COVID-19 pandemic and, 114
Thames (river), 23, 28
theory of pragmatism, 55
thermostatic institutions, 122
The Three Body Problem (Cixin), 144
Thunberg, Greta, 74
time: horizons of, 138–39, 167; power and, 140; qualitative approaches regarding,

INDEX 241

140–41; time preferences, 24; time revolution, 193
Toffler, Alvin, 160, 196, 198, 210n6
trade liberalization, 31
transition theory, 153–54, 155
transnational governance, 3, 65. *See also* governance
Tremmel, Jörg, 88, 119
triggers, 122–27, 167, 173–74
Trump, Donald, 111–12
trusteeship: authority of, 82; California Air Resources Board (CARB) as, 84–86; central banks and, 82, 83; challenges regarding, 87, 126; courts/judges as, 89; creating, 81; decision-making power of, 83; defined, 79; delegation to, 79, 84; dilemma regarding, 79; empowerment of, 89–90; examples of, 86; influence in, 94; legal long-termism and, 88–90; mandates for, 172; needs for, 81–82; orientation of, 81–82; overview of, 78–90, 167; power to, 172; trust in, 84
Tunisia, 75

UK Climate Change Committee (UKCCC), 74–75, 77–78
uncertainty: in climate change, 39; early action paradox and, 37–42; experimentalism and, 54–60; of long problems, 39; pandemic disease and, 40; problem length and, 38; as source of limitations, 38; technological development and, 41
Under 2 Coalition, 110
UN Environment Programme, 96
United Arab Emirates, 75
United Kingdom (UK): Better Business Act, 92; Climate Assembly UK, 97; net-zero transition plan requirement in, 93; one in, two out rule in, 116
United Nations (UN), 13, 34, 35, 170
United Nations Educational, Scientific, and Cultural Organization (UNESCO), 96
United Nations Framework Convention on Climate Change (UNFCCC), 34, 36–37, 57, 60, 104–5, 108, 119, 120, 179

United States: Clean Air Act of, 80; climate change activism in, 100; climate-related financial risk requirement in, 93; Declaration of Independence of, 193; and Green New Deal, 170; Paris Agreement and, 111–12; social security payments in, 125–26; Soviet Union and, 120; sunset clauses and, 116–17; wealth funds in, 124; in World War II, 143; zoning in, 117
UN Security Council, 120
urban planning, successive limited comparisons and, 39
Urgenda Foundation v. State of the Netherlands, 87
Urpelainen, Johannes, 133

Venkataraman, Bina, 44
Vesuvius, Mount, 44–45
veto players, 104, 115
Vickers, Geoffrey, 12, 17, 181, 197
Vickers, John, 169
Victor, David, 54, 56, 136
Vietnam War, 120
Vision 2030 plan (Saudi Arabia), 107, 125

war, 120, 123, 129–30, 143, 162–63, 195; as short problem, 130
We Are Still In coalition (United States), 111–12
webcraft, for climate change, 65
Wellbeing Economy Alliance, 92
Wells, H. G., 157, 158
Welsh Commission for Future Generations, 75
Wenger, Andreas, 158
white supremacy, 7
wicked problem, 200n26
World Bank, 83, 121
World Meteorological Organization, 96
World War II, 143

Xi Jinping, 109

Young, Oran, 103–4, 106
youth, activism of, 100–101

zoning, 117